HIGHLIGHTS OF BRITISH SCIENCE

HIGHLIGHTS OF BRITISH SCIENCE

Based on the subjects of exhibits arranged
for the Jubilee Exhibition at the Royal Society
20 to 25 June 1977

LONDON
THE ROYAL SOCIETY
1978

Printed in Great Britain for the Royal Society
by
Unwin Brothers Limited
The Gresham Press, Old Woking, Surrey

ISBN 0 85403 104 9

Published by the Royal Society
6 Carlton House Terrace, London SW1Y 5AG

CONTENTS

PREFACE

By long tradition the Royal Society holds two conversaziones each year, in the months of May and June, and at them interesting new scientific discoveries and developments made by Fellows and their associates are on display. To mark the Silver Jubilee of Her Majesty the Queen's accession to the throne the Society's Council decided to mount a special exhibition in 1977 having as its theme scientific developments in the United Kingdom during the Queen's reign. The exhibition was held in the Society's rooms from 20 to 25 June and was based substantially on exhibits prepared by Fellows for our conversaziones, one of which was indeed held on the evening of 23 June during the period of the exhibition. The exhibition was well attended and most successful. The interest which it aroused among visitors suggested to Council that some of the achievements of modern science and, in particular, the part played in them by British scientists were less well-known than they ought to be and deserved a wider publicity. This volume is an attempt to fill this gap.

Britain has a great tradition in natural science. From the time of the so-called scientific revolution of the mid-seventeenth century with the foundation of the Royal Society at the heart of it, British scientists have been in the forefront of advances not merely in fundamental science but also in its application to the practical problems of everyday life. Science, of course, knows no national boundaries and its owes much of its progress and vitality to the free exchange of information and ideas between scientists in all countries. It is nevertheless true that, for a variety of reasons, some countries, of which Britain is one, have made contributions out of all proportion to their physical size and population. Today we are in the midst of a worldwide economic recession which has indeed affected Britain more severely than some other highly industrialized nations. Since economic progress and with it our living standards depend nowadays almost entirely on advances in science and in technology based upon it, it is perhaps not surprising that some members of the public should wonder whether we have fallen behind other nations and whether our expenditure on science and scientific research is being misdirected. Nothing could be further from the truth. Our record during the past twenty-five years is an enviable one and our scientific research is vigorous as ever, flushed with success and full of promise. Britain has made and continues to make outstanding contributions in many and diverse fields of science. Some of these are set out in this book and it is my hope that its contents will not only stimulate appreciation of some highlights in British science but will indicate also its promise for the future. Our warmest thanks are due to all the contributors for the excellent essays they have written and to Dr Christine Stickland for all her work in preparing this volume for publication.

DISCOVERIES ABOUT THE UNIVERSE

R.L.F. BOYD, F.R.S.

Mullard Space Science Laboratory of University College London

Discoveries made by British scientists in the realm of astronomy have been brought about in many cases by theoretical insight which has a long history in United Kingdom astronomy and cosmology, but almost without exception they have been made possible by technical developments and improvements in which our scientists and engineers have played a pioneering or major rôle. While it is appropriate here that we emphasize the British rôle it is proper also to stress that astronomy is one of the most international of disciplines and that much work has been collaborative. Parallel progress has been made in many places outside our Commonwealth. In particular our enviable progress in astronomy from spacecraft has only been possible because of generosity on the part of the United States through its National Aeronautics and Space Agency.

The last quarter of a century has seen a change in attitudes to astronomy which justify the use of the word revolutionary as it derives from 'De Revolutionibus Orbium Caelestium'. The impact of these attitudes and the techniques with which they have grown up bid fair to be as epoch making as that. No longer is astronomy thought of as a collection of different disciplines—radio astronomy for physicists with an amateur interest in real astronomy; real astronomy to be conducted with appropriate mortifying of the flesh by long vigils with strained eyes in freezing observatories; theoretical astronomy and cosmology for speculating scientists with a passion for mathematics and inadequate data to constrain them. Today all combine to catch and consider each energy-information-bearing photon, whether it arrives through the radio window in our atmosphere with wavelengths from less than a millimetre to many metres, or through the many infrared windows scattered in narrow bands from submillimetric wavelength to micrometres, or, still of undiminished importance, through the familiar optical window.

To the ground-based observations are added those made from balloons, rockets and spacecraft in the infrared, ultraviolet, X-ray and γ-ray bands with further information from the study of cosmic rays, meteors and meteorites along with instrumental or manned visits to solar system bodies. Even neutrinos and gravity waves are sought.

Although this account must be drastically and arbitrarily selective, the last twenty five years in UK astronomy is a story of technical innovation in many of these areas. Our radio astronomy aerials are national monuments in themselves

1

(figure 1, plate I) but the 150 inch Anglo-Australian optical telescope at Siding Spring in Australia with its associated Schmidt camera are less evident and less well known and our ultraviolet and X-ray instruments have not been seen since their launching into orbit though the latter are still sending back their often unique observations.

Of the new astronomies that depend on spacecraft to carry the instruments above the atmosphere X-ray astronomy in the UK has been amongst the world leaders since the first cosmic X-ray reflecting telescopes to go into orbit were built by the Mullard Space Science Laboratory of University College London and carried aloft on the American satellite Copernicus. Copernicus and the British-built Ariel V satellite, basing their studies largely on the superb survey work of the American Uhuru satellite, carried X-ray astronomy from the exploratory stage to the study of chosen objects. Quite as important to the progress of UK astronomy has been the establishment, in collaboration with Australia, of the Southern Hemisphere Observatory at Siding Spring. This Observatory has two instruments second to none in performance, excellently situated for study of the formerly rather neglected southern sky, with its view of the galactic centre and of our nearest neighbouring galaxies, the Magellanic Clouds. In the northern hemisphere valuable experience has been gained with the 100-inch Isaac Newton Telescope, inaugurated by the Queen at Herstmonceux in 1967 and soon to be moved with a new mirror to a lower latitude site where there will be much better atmospheric conditions; it is hoped that a 4.5 m giant telescope will follow. This progress in optical astronomy has taken place alongside huge developments in computer power which enable automatic corrections to be made for the minute distortions that occur as the scores of tons of telescope changes its attitude. On-line computers also enable accurate control from a comfortable console, of the telescope and all of its systems. Of comparable significance has been the development of detection, data processing and recording techniques which replace the photographic plate with its many problems and poor photon efficiency by the perfectly linear, highly sensitive response of photocathodes.

In what follows we shall look at a couple of theoretical fields to which British workers have made outstanding contributions and then describe the main developments in X-ray, Ultraviolet, Visual and Radio astronomy and some of their contributions to a number of important discoveries about the Universe.

THEORETICAL ASTRONOMY

British and Commonwealth theorists have played a large part in the construction of the comprehensive picture that we now have of the origin and evolution of the Universe and its constituents. On conventional ideas the recognition that the Universe is expanding would imply that its density was higher in the past. However, in 1948 Bondi, Gold and Hoyle proposed an alternative 'Steady State' theory: as the galaxies moved apart new matter was created and new galaxies evolved so that the Universe looked the same at all times and the awkward problem of having a beginning was avoided. Even though this attractive theory turned out to be incorrect in the end, it inspired a lot of important theoretical and observational work including work by Ryle and his group which showed that the density of radio sources, and hence presumably of galaxies, was higher in the past than it is now. The whole Universe is evolving. The question then arose: if one went back in time would the density of the Universe become infinite at some moment, the 'Big-Bang' theory, or would there have been a period of large, but finite, density, before which the motion of the galaxies was reversed, the 'Bounce' theory? It is fairly easy to show that if Einstein's General Theory of Relativity is correct and if other galaxies are moving exactly in the direction away from us, then about ten thousand million years ago all the galaxies would have been on top of each other and the density would have been infinite. However, one would expect that the galaxies would have some random transverse velocity component causing them to miss each other and to have given a 'bounce' instead. Penrose and Hawking in a series of papers between 1965 and 1970 showed that this was not the case: if general relativity is correct there must have been a 'singularity' in the past where the density of matter and the curvature of space–time were infinite. The singularity would constitute a beginning to the Universe, a point of Creation. This 'Big-Bang' picture is strongly supported by the discovery of a background microwave radiation which is interpreted as a relic of the immensely hot and dense early stages. As the Universe expanded the radiation would have cooled until it fell to the 3K that we observe today.

In the very hot conditions to start with only the hydrogen would have been present but as the Universe expanded and cooled, about a quarter of the hydrogen atoms would have combined to form helium and a small amount of deuterium (heavy hydrogen). These theoretical abundances of helium and deuterium are in very good agreement with what we observe today.

Although the Big-Bang theory was very successful with helium and deuterium the problem of the formation of the other elements remained because there are no stable nuclei of atomic weights 5 or 8, so the Big-Bang could not account for the elements which have an atomic weight greater than 4. So, how did they arise? The solution was found in 1953 by Hoyle who at that time still adhered to the Steady State theory. It was known that normal stars transformed hydrogen into helium as the source of their energy. In the Big-Bang the absence of a stable nucleus with atomic weight 8 prevented two helium nuclei combining, but a star exists long enough for three helium nuclei to collide at the same moment to give a nucleus of carbon (atomic weight 12) if the carbon nucleus had a resonant excited state at just the right energy. Hoyle persuaded Fowler of Caltech to look for the resonance and sure enough it was there.

Once the hurdle of the missing nucleus at mass 8 was overcome, it was fairly easy to explain the formation of the elements up to iron but to explain the heavier elements required a new process because iron is the most tightly bound of all nuclei. In a classic paper Margaret and Geoffrey Burbidge, Hoyle and

4 R.L.F. BOYD

Fowler showed how these elements were built up from iron by neutron bombardment in supernova explosions. These same explosions scattered the newly formed heavy elements back into the interstellar medium to form the next generation of stars and planetary systems. This historic paper is usually referred to as B²FH.

Calculations of stellar evolution based on the B²FH paper have indicated that most stars with masses between about five and ten times the mass of the Sun probably explode as supernovae leaving behind a very highly compressed remnant of about one solar mass but a radius of only about 10 kilometres, composed mainly of neutrons packed together as tightly as in an atomic nucleus. Such neutron stars occur in pulsars and pulsating X-ray sources. Similar calculations indicate that, when it has exhausted its nuclear fuel, a star of more than about ten solar masses would not explode. Instead, unable to support itself against its own gravity it would shrink to a radius of about 30 kilometres when its gravitational field would become so strong that it would drag back any further light emitted by the star. Such invisible objects which still exert the same gravitational pull as the star which collapsed are known as Black Holes. The concept was first proposed by Laplace in 1798 but no one took it very seriously until about 1965. Since then a large part of the work on black holes has been carried out by British and Commonwealth scientists.

In 1971 Rees and Pringle suggested that if one had a black hole in orbit around a normal star, material of the star might be dragged or blown off and fall into the black hole. As it did so it would develop a spiral motion and would get very hot emitting X-rays. A year later the American satellite Uhuru discovered a rapidly fluctuating X-ray source Cygnus X-1 since extensively studied by our instruments on Copernicus and identified by the Royal Greenwich Observatory as a binary system consisting of a normal star and an unseen non-luminous object with a mass of about ten solar masses. This is a strong black hole candidate.

The results of Penrose and Hawking imply that inside the black hole the star continues to collapse until it forms a singularity of infinite density and space–time curvature. This singularity like the one in the Big-Bang is an 'edge' to space–time, in this case an 'end' rather than a 'beginning' at least for the material of the star. At such an edge the valid laws of physics are unknown.

Outside the black hole the gravitational field would rapidly settle down to a stationary state which depended only on the mass, angular momentum and electric charge of the star that collapsed but was independent of all its other properties such as shape, constitution or magnetic field. This result, known as the theorem: 'A black hole has no hair', was established by the work of Israel, Carter, Hawking and Robinson. It shows that a very large amount of information about the star is irretrievably lost down the black hole in a gravitational collapse. Because loss of information about a system is equivalent to an increase of disorder, this suggests that black holes may possess the thermodynamic quantity, entropy, which is a measure of the disorder or randomness of a system. In fact it had been discovered that black holes have a quantity very analogous to entropy: namely, the surface area of their boundaries. This would increase whenever more matter or radiation fell into a black hole and, when two black holes collided and merged together to form a single black hole, the surface area of the final black hole would be greater than the sum of the surface areas of the original black holes.

There was, however, an apparently insurmountable obstacle to endowing a black hole with a finite entropy because this would imply some finite temperature and should, according to thermodynamics, be able to remain in equilibrium with thermal radiation at the same temperature. But this would not be possible if, as was generally thought, nothing could escape from a black hole, for the black hole would absorb some of the thermal radiation but would not be able to emit anything in return.

The paradox was resolved in 1974 when Hawking discovered that when the quantum mechanical Uncertainty Principle was taken into account, black holes did indeed emit radiation with a thermal spectrum. One can think of the radiation as arising in the following way: the Uncertainty Principle implies that energy cannot be exactly defined over very short periods of time. This means that even in a vacuum, the lowest energy state, it is possible for pairs of particles and anti-particles spontaneously to appear together, move apart and then come together again and annihilate each other within the time interval prescribed by the Uncertainty Principle. These particles are said to be 'virtual' because, unlike 'real' particles, they cannot be observed directly, though their indirect effects have been measured. When a black hole is present, one of the particles of a pair may fall into it leaving the other without a partner with whom to annihilate. It may follow its mate into the black hole or it may escape, appearing to have been emitted from the black hole. In this way the black hole emits particles and radiation just as if it were a hot body with a temperature inversely proportional to its mass. For a black hole of solar mass, the temperature is only about 10^{-7} K so the radiation would be negligible and it would, to all intents and purposes, be completely black. However the Universe might also contain very much smaller black holes formed by the collapse of small regions in the very hot and dense medium following the Big-Bang. Such 'primordial' black holes (the size of an elementary particle), might have masses of 10^9 tons. They would have a temperature of about 10^{11} K and would be emitting about six thousand megawatts as gamma rays. As they emitted energy they would get smaller and hotter and eventually disappear completely in a tremendous explosion equivalent to millions of H-bombs.

In the last couple of years there have been several attempts to detect primordial black holes either by their steady gamma-ray emission or by their explosions. So far only an upper limit of about two hundred per cubic light year has been obtained. If more sensitive measurements do not discover any primordial black holes, that will still be very important because it will indicate that the material in the early Universe must have been very smoothly distributed.

X-ray Astronomy

It was natural that the huge engineering achievements in rocketry started during the war especially in Penemunde, and carried on with such vigour ever since, should eventually be put to less belligerent use. So it was that during the nineteen fifties the United States began to deploy astronomical instruments above the absorbing atmosphere and to open up a new window in the electromagnetic spectrum. The same period saw the development of the Skylark research rocket in the United Kingdom and its use at Woomera initially for upper atmospheric studies, and in 1961 on the first survey of the southern sky in the ultraviolet. About this time the Sun was being studied in the U-V and X-rays—the first and only known X-ray source at that time—and plans were being made both in the US and the UK to look for other, cosmic, X-ray sources. Such a source was discovered in 1962 by Giacconi, working in the States, and in the succeeding eight years about fifty more sources were found by astronomers throughout the world. Although these were found with increasingly sophisticated and sensitive instruments, the instruments were flown on rockets which had useful observing times limited to about five minutes per flight. During this period the US was planning and building a satellite (Uhuru) launched in 1970 and devoted to a sky survey, and in the UK plans were made, even before any cosmic sources were known, for the use of reflecting telescopes for the study of individual objects. The latter resulted in the set of reflecting X-ray telescopes launched on Copernicus in 1972. An unexpected delay in the schedule of this US Orbiting Astronomical Observatory turned out to be fortunate as its very accurately pointed ensemble of telescopes could be deployed on a map of the X-ray sky now provided by Uhuru which had already trebled the number of candidate objects. Two years later the UK satellite Ariel V was launched, devoted wholly to X-ray astronomy, with the capacity both to make a scanning survey and to point to selected objects. It could determine positions to 1 min arc—very important in seeking identifications with radio or optical counterparts—and had, furthermore, unprecedented resolution and coverage in photon energy. At the time of writing (November 1977) Copernicus and Ariel V are still operating and have provided many, perhaps most of the recent advances in observational X-ray astronomy.

A glance at figure 2 will show that the X-ray sky known in 1968 contained 34 sources, almost all unidentified with astronomical objects and mostly concentrated along the galactic equator and therefore presumably members of our Galaxy. Following the Uhuru and Ariel V surveys the number of objects detected approached 300 with nearly 100 being extragalactic of which more than half have been identified. A smaller proportion of galactic objects has been identified.

X-ray astronomy has from the first been carried out with proportional gas counters as photon detectors while since the pioneering work on Copernicus there has been an increasing trend towards the use of grazing incidence mirrors which for grazing angles of the order of a degree enable a paraboloid to focus

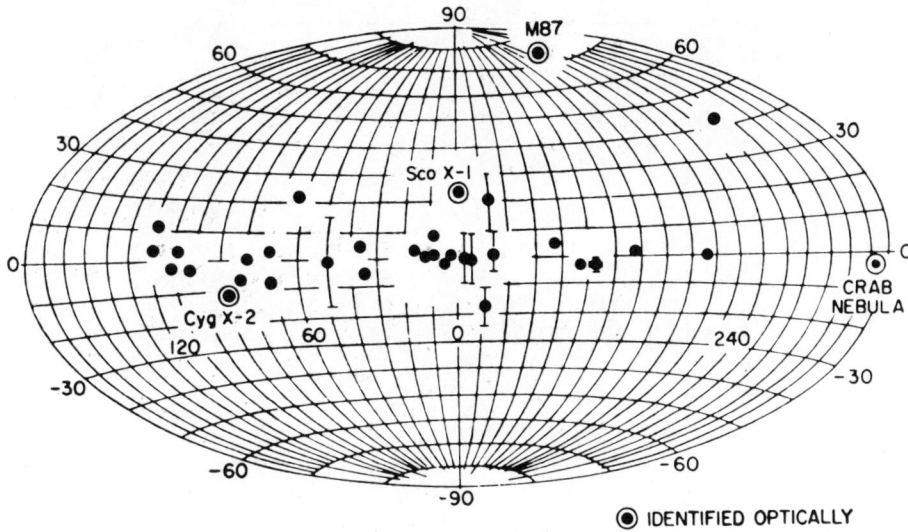

FIGURE 2. The X-ray sky in 1968, before the advent of X-ray astronomy satellites. (*From* Giacconi *et al.*, *Ann. Rev. Astr. Astrophys.*, **6**, 373, 1968.)

on-axis X-rays. With two reflections, the second from a hyperbola, imaging an extended field is possible. Position-sensitive X-ray detectors for recording these pictures have been developed in British universities and these and other devices are being used in an on-going programme of X-ray astronomy involving UK-6 and the European Space Agency satellite EXOSAT.

One of the earliest indications of what might be expected of cosmic X-ray sources came, naturally, from the behaviour of the Sun. Indeed in 1962 Ariel I showed that during a solar flare emission from the tiny area of the flare swamped the entire output of X-rays from the Sun's outer atmosphere (corona). A consideration of the Sun (which as an X-ray star is so weak that it would be barely detectable with existing techniques if it were as distant as the nearer stars) suggested that if they were to be observable cosmic X-ray sources would need to be much brighter in X-rays and that probably other mechanisms than those operating on the Sun would be involved both as energy sources and radiation processes. In the event the very first X-ray source discovered, Scorpius X-1, was so unexpectedly energetic (it gave out in X-rays alone 10^3 times as much energy as the total solar output) that an energy source different from the stellar nuclear furnace was clearly involved. We now know that the heating of matter by its falling into deep gravitational potential wells is the most important energy generating mechanism and the radiation processes have turned out to be thermal emission from hot plasmas, synchrotron radiation (so important in radio sources) and inverse Compton scattering of relativistic electrons by the cosmic microwave background photons. Thus it is that X-ray astronomy has become a prime diagnostic technique for the study of such condensed objects as black

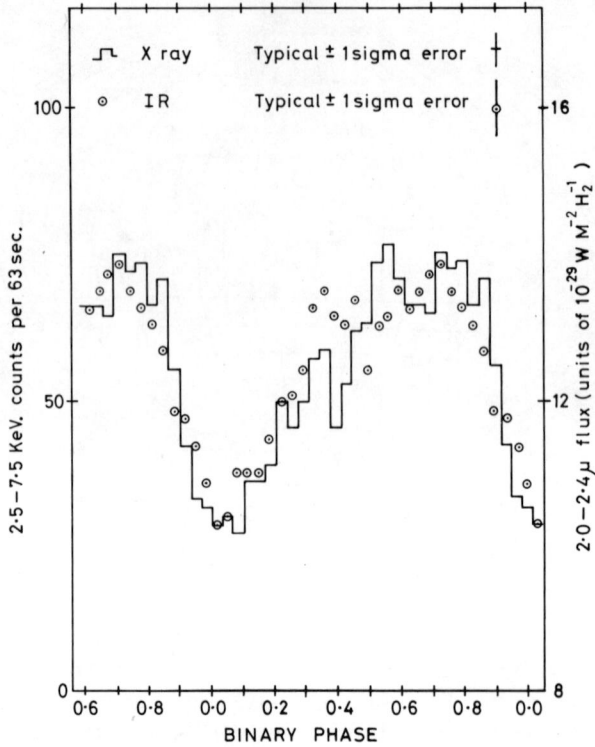

FIGURE 4.   X-ray and infrared signals from Cyg X-3 showing the same pattern of modulation. This provides convincing evidence that the X-ray and infrared sources are associated. (*From* Mason *et al.*, *Astrophys.J.*, **207**, 78, 1976, by permission of the *Astrophysical Journal* and the University of Chicago Press.)

holes, neutron stars, white dwarfs, the nuclei of active galaxies and very probably quasars and the potential wells associated with globular clusters and clusters of galaxies. A little less directly it is associated with the rôle of gravity as a very important technique in the study of the energy release during the gravitational collapse of a dying star which results in a supernova explosion and in the study of those very closely bound stellar pairs known as symbiotic binaries.

Most optically observed astronomical phenomena proceed on a fairly leisurely time scale but, just because X-ray astronomy is so often concerned with relatively small amounts of matter at temperatures of many millions of degrees radiating huge energy fluxes, cosmic X-ray sources can be expected to vary within a millisecond and the X-radiation from whole galaxies can be associated with intense sources whose dimensions are less than a light year as evidenced by variations occurring in times much less than one year.

FIGURE 1. The Mk IA radio telescope at Jodrell Bank. The telescope, previously known as the Mk I, was completed in 1957; it was modified, upgraded (and renamed) in 1971. It has a diameter of 250 ft (76.2 m), and was for many years the largest fully steerable radio telescope in the world. (*Courtesy* Nuffield Radio Astronomy Laboratories, Jodrell Bank.)

PLATE II (BOYD)

FIGURE 3. The Ariel V satellite; number 5 in the series of UK satellites, it is dedicated to X-ray astronomy. It was launched in October 1974 and has proved successful beyond all expectations. It carries a variety of X-ray telescopes provided by British universities. (*Courtesy Appleton Laboratory.*)

FIGURE 10. The Anglo-Australian telescope. The 3.9-m primary mirror is housed at the centre of the horse-shoe and the optical axis is pointing top right. The control console is behind the lighted window (centre back). (An Anglo-Australian Telescope photograph.)

PLATE IV (BOYD)

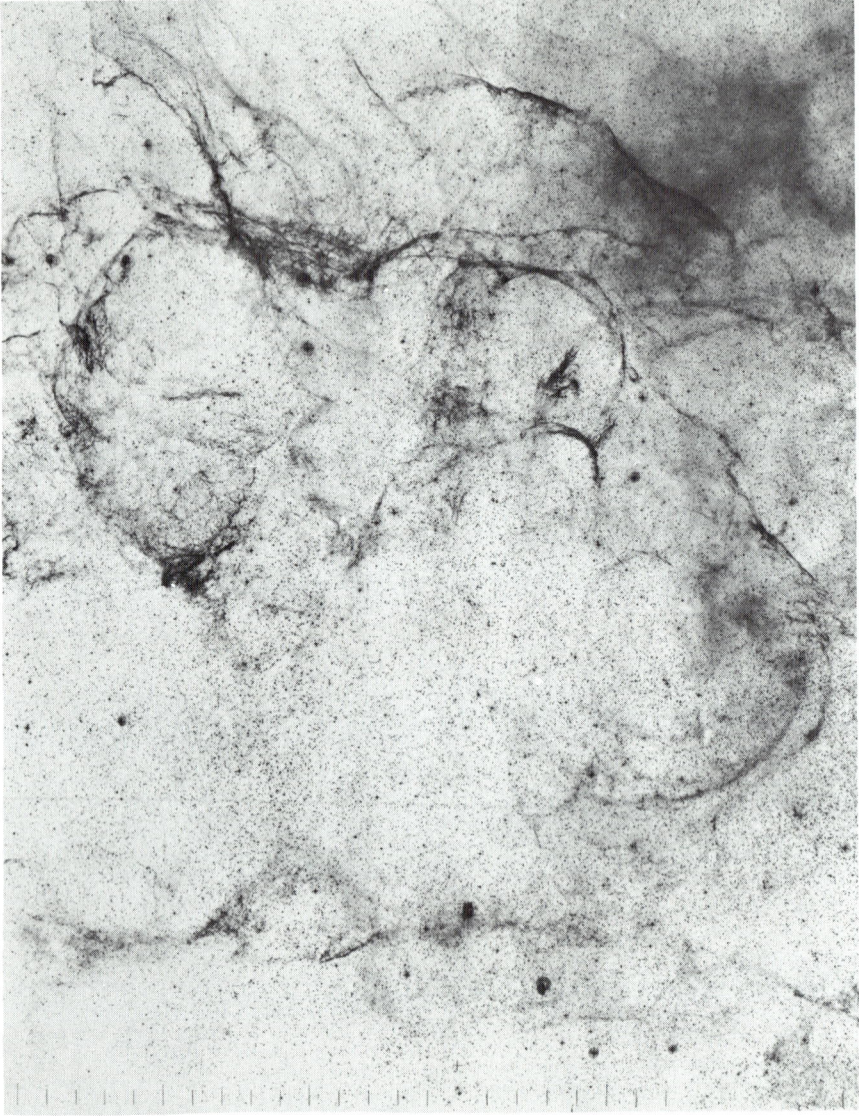

FIGURE 11. A negative photograph of the area of sky around the Vela pulsar taken on the UK Schmidt camera with a narrow band interference filter. The features which appear as dark arcs are great loops and bubbles of gas which were blown off in one or more supernovae explosions 10 000 years ago. The thousands of black specks are faint stars, many of them brighter than the Vela pulsar. (*Courtesy* Royal Greenwich Observatory.)

Because Copernicus and Ariel V (figure 3, plate II) have, in the seven years since the launch of Uhuru, been almost alone in their ability to point at X-ray sources selected for their astrophysical interest, the UK has been uniquely placed to follow up the many questions raised and challenges presented by the survey conducted by Uhuru. (A successful Dutch satellite ANS, launched two years after Copernicus, also carried pointed X-ray telescopes but its observing life was rather short.) Ariel V is also equipped with sky survey facilities which scan great circles of sky at right angles to the spin axis of the satellite. This sky survey instrument early established the transient nature of many sources so that the Ariel Survey not only added some fifty new sources to the earlier survey but showed that some of the Uhuru sources had faded away.

The first of the transients discovered by Ariel V A1542-61 was found to lie close to the 17th magnitude star which subsequently faded in a similar manner, thus confirming the association. This illustrates one of the ways of carrying out identification of sources, by a coincidence in temporal (secular or regular) behaviour. Another example is illustrated in figure 4 where Copernicus observations are seen to make possible the identification of Cygnus X-3 with an infrared source, this time by the regular periodicity of both. The more direct approach to the important problem of identification is accurately to determine the position of both the X-ray source and its candidate source in another part of the spectrum. Copernicus and Ariel V could achieve a precision of about 1 min arc but substantially better accuracy has been obtained with Copernicus from time to time using the method of lunar occultation.

One of the transients discovered in the constellation of Centaurus just before Christmas 1974 naturally earned the nickname Cen X-mas. Its correct nomenclature is A1118-61 and its transient nature is shown clearly in the light curve of figure 5. Later it was found to be also a member of another class of

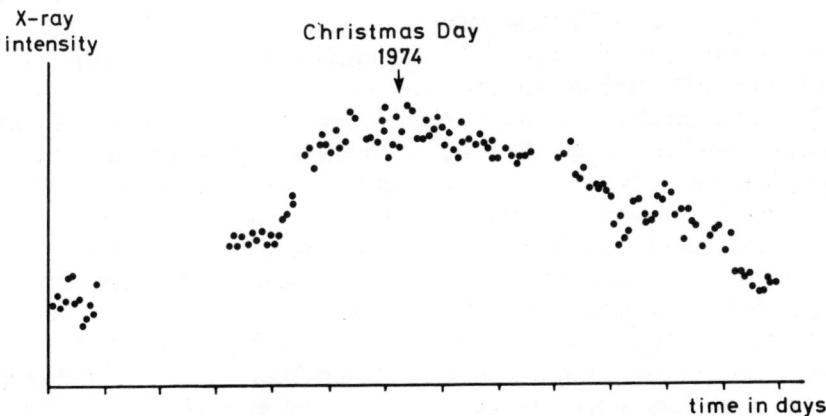

FIGURE 5. The flaring and fading of Cen X-mas, so named because it reached peak brightness on Christmas day. One of the outstanding features of the Ariel V observations has been the detection of a number of remarkably bright but short-lived X-ray sources.

X-ray
intensity

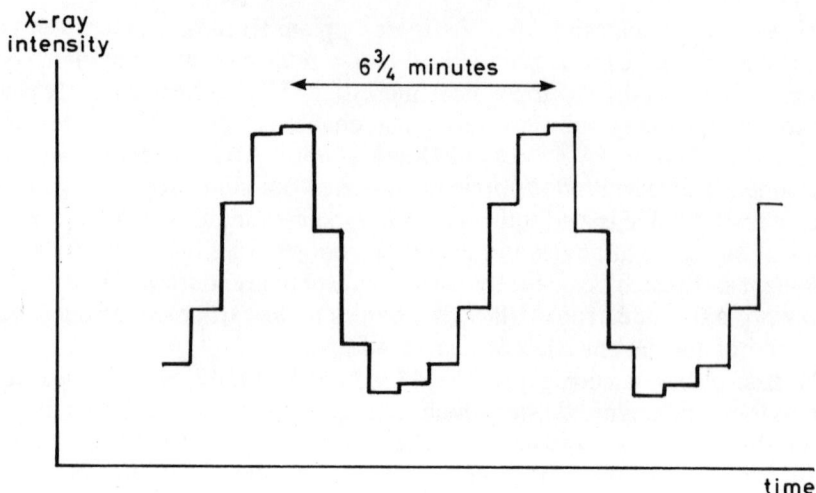

time

FIGURE 6.   As well as the overall brightening and fading of Cen X-mas, the X-ray emission was
found to be pulsed with a period of 6¾ minutes. The emission from the star is believed to be beamed
and the slow flashing is caused by the beam sweeping round and round as the star rotates every 6¾
minutes.

objects identified by Ariel V and Copernicus. These are the Slow Rotators. They
are really X-ray pulsars (see figure 6). Like radio pulsars, X-ray pulsars (or
pulsators) are believed to be neutron stars; but unlike radio pulsars these neutron
stars are not single stars, but occur in binary systems as partners to normal stars.
They are normally only visible in X-rays.

Also possibly containing neutron stars are the supernova explosion remnants
which are often strong radio sources and weakly observable in the visible. The
arc minute resolution of the Copernicus telescopes made it possible for the first
time to map a number of these remnants in the X-ray band (see figure 7) and
contributed substantially to our understanding of the behaviour of the interstellar
shock wave generated by supernovae.

The search for clear evidence of black holes is an exciting aspect of present
day astronomy. Here again the UK instruments have played an important part.
Copernicus has studied the prime candidate Cygnus X-1 mentioned earlier and
confirmed its association with the B0 supergiant HDE226868 by many
observations of fading effects related to the period of the binary.

Quite as important as the discoveries within our Galaxy have been those
beyond. Some fifteen Seyfert galaxies with their highly agitated cores have been
identified as X-ray sources and three new (i.e. formerly unknown as such)
Seyfert galaxies have been discovered from the Ariel V survey. These systems
are not well understood but their place in the evolution of galaxies and
relationship to quasars and possibly black holes is of great cosmological
significance. It is in this context that the evidence that some are X-ray sources is
so important.

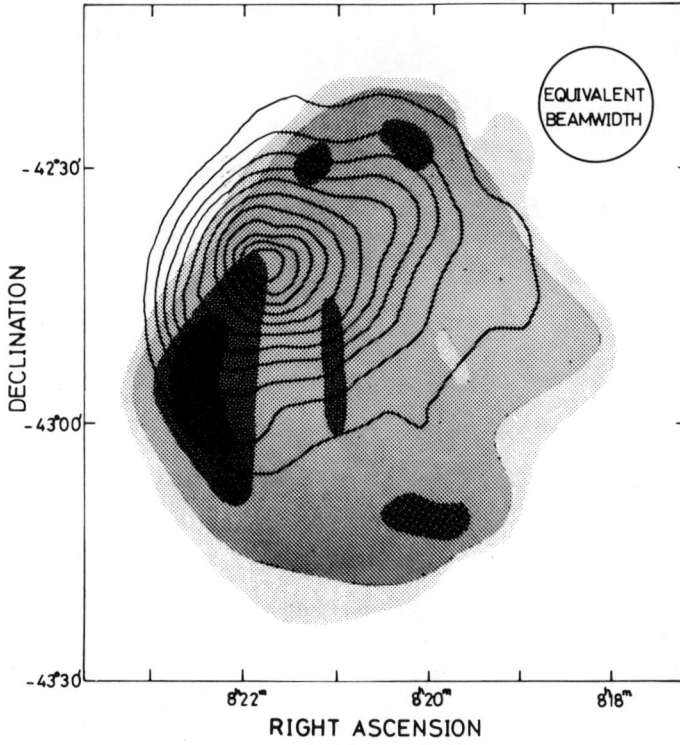

FIGURE 7. Contours of X-ray brightness ($\frac{1}{4}$–$1\frac{1}{2}$ keV) of the supernova remnant Puppis A, superimposed on a radio brightness map. (*From* Culhane, *Vistas in Astronomy*, **19**, 1, by permission of Pergamon Press Ltd.)

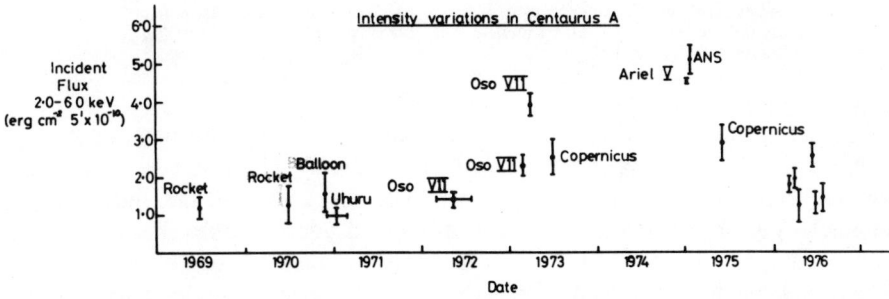

FIGURE 8. Changes in the X-ray brightness of Centaurus A. The first properly authenticated case of a variable, extragalactic X-ray source. The variability was first noticed in 1973 in Copernicus data and subsequently found in earlier, American observations. (*After* Stark *et al.*, *Mon. Not. R. astr. Soc.*, **174**, 35P, 1976, by permission.)

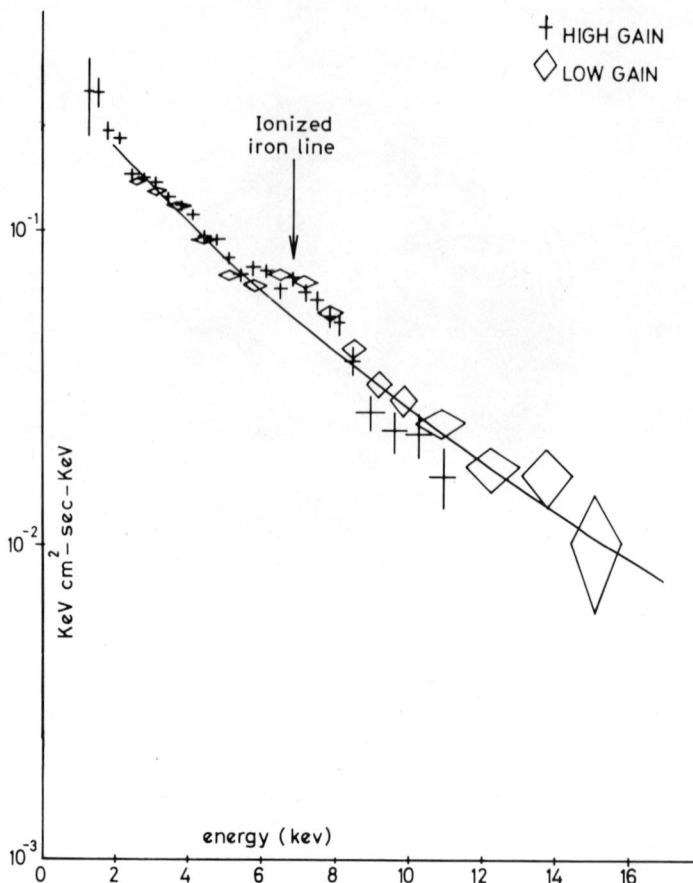

FIGURE 9. The spectrum of the X-ray emission from the Perseus cluster of galaxies. As well as containing some 500 galaxies this cluster contains a hot, but tenuous gas between the galaxies, which is responsible for most of the X-ray emission. The bump in the spectrum at about 7 keV is X-ray line emission from highly ionized iron lines. (*After* Mitchell *et al.*, *Mon. Not. R. astr. Soc.*, **176**, 29P, 1976, by permission.)

Another unique discovery made by Copernicus is the variability of the X-rays from the giant radio galaxy Centaurus-A (see figure 8). The whole system is well over a million light years long and the bright optical core is 40 000 light years in diameter. Variations of the X-rays on a scale of years implies that they come from a tiny part observed to be near the centre. Again the idea of an accreting gigantic black hole is raised. Even the space between galaxies about which almost nothing was known until recently, has been found to radiate X-rays from highly ionized, very hot iron atoms (see figure 9 for evidence of this in the X-rays from the Perseus cluster of galaxies). The source, heating and quantity of this intracluster matter is also of great cosmological importance.

The United Kingdom has a continuing programme of X-ray astronomy. The satellite UK-6 is due for launch early in 1979. The UK is involved in all of the experiments on the European Space Agency Satellite EXOSAT (due early 1981) and in X-ray astronomy experiments on Spacelab I and II.

ULTRAVIOLET ASTRONOMY

Ultraviolet radiation like X-radiation is blocked by the Earth's atmosphere and all observations shorter than 3000 ångströms call for balloons, rockets or satellites. We have already mentioned that the first survey of the Southern Sky was made from Woomera by a United Kingdom Skylark rocket. These rockets together with stabilized balloon platforms have continued to make important contributions, but the greatest progress has come from the Princeton University U-V spectrometric telescope on Copernicus and the UK telescope in the ESRO TD-1 satellite. This latter experiment scanned the whole sky and detected about 30 000 stars down to 9th visual magnitude. It provided low-resolution spectra in absolute flux units between 1350 Å and 2550 Å, and a catalogue describing 1400 of the brighter stars in both numerical and graphical form has been published. A second catalogue giving just four broad-band fluxes for 30 000 stars is being prepared.

Probably the most exciting of several important discoveries made with this instrument concerns the size of the population of sub-dwarf stars. About 1000 of these stars have been observed. They are relatively faint and appear to have very little interstellar reddening. Considering the strong effect of interstellar extinction in the ultraviolet, one can conclude that they are relatively close to us and that their low apparent brightness implies that they are subluminous by comparison with main-sequence stars—whence their name 'sub-dwarf' stars. They occupy a region of the Hertzsprung–Russell diagram (a graph of luminosity against temperature valuable in classifying stars and interpreting their evolution) which has always been thought to have a very sparse population, but these new ultraviolet observations indicate that the population of stars in this region is very large and may in fact be comparable with that of the main sequence.

A typical hot sub-dwarf is the one designated BD + 39°3226. Formed in the galactic halo, where stars generally are built from primordial hydrogen and helium, it displays no evidence for hydrogen in its outer layers, which are almost pure helium. This object must therefore be highly evolved, its initial store of hydrogen having been converted by nuclear fusion into helium. Though similar in size to the Sun, BD + 39°3226 is eight times hotter.

[Since this was written UK scientists have been involved in a further remarkable success in this field—the launch of the International Ultraviolet Explorer.]

OPTICAL ASTRONOMY

An agreement between the participating Governments to build the Anglo-Australian telescope (AAT) was reached in April 1967. The telescope stands on Siding Spring Mountain, New South Wales, near Coonabarabran where there is roughly twice as much clear night sky and much better seeing conditions than at the best sites in the United Kingdom. The telescope (figure 10, plate III) was inaugurated by the Prince of Wales in October 1974. Its 3.9-m diameter primary mirror is made from the glass–ceramic material Cervit having a negligible coefficient of expansion at working temperatures which is of great importance because with a mass of 15 tonnes its thermal relaxation time is several days. The telescope whose moving parts weigh 250 tonnes and the dome weighing 500 tonnes are controlled by computer which routinely points the instrument anywhere in the sky to an accuracy of 2.5 arc seconds using an algorithm which takes account of all the small terms in the astronomical reference frame together with the flexure and residual misalignment of the telescope's mounting. The telescope can be used with focal ratios of f 3.26, 8, and 15 and is equipped with a variety of auxiliary equipment and detectors. In addition there is an f/36 coudé system. One of the most frequently used instruments is an intermediate dispersion spectrograph whose output is recorded by the Image Photon Counting System described later. It is also used in twilight and daylight to study cosmic infrared and microwave sources.

The observation programmes are chosen for their scientific interest and because they exploit the telescope's unique performance; many are related to the cosmic sources of X-radiation being studied by Ariel V and Copernicus. Others are concerned with quasars and galaxies which are strong sources of radio waves.

The UK 1.2-m Schmidt Telescope was originally planned to complement the Anglo-Australian telescope under construction on the same site, but has turned out to be a powerful research instrument in its own right. The Schmidt design provides a wide field for direct photography, ideal for sky surveys. The current primary task of the UK Schmidt is to produce a photographic atlas of the southern skies, which involves taking a set of 606 overlapping photographs under the very best conditions of astronomical 'seeing'. This survey reaches very much fainter and more distant stars and galaxies than any previous survey of the southern sky and indeed, by taking advantage of recent advances in photographic and telescope technology, goes considerably deeper into space than the corresponding survey of the northern sky made by the similar Palomar 48-inch Schmidt during the early 1950's. To permit the telescope to focus accurately light of all wavelengths (ranging from 320 nm to 1000 nm), it has recently been fitted with a cemented doublet achromatic corrector lens; with a diameter of 1.2 m, this is believed to be the largest cemented doublet ever made.

The real forte of the instrument is its ability to photograph nebulosity at very low surface brightness, the limit for detection being only about one per cent of the overall sky brightness, and this ability has led to many new discoveries such as extended planetary nebulae, large numbers of previously uncatalogued galaxies, and faint extensions to known galaxies which have sometimes led to a radical revision of their classification. Recently, a dwarf member of our Local Group of galaxies has been discovered in the constellation of Carina.

Two advances should be mentioned which depend on technical innovations. Meaburn of Manchester University has designed a mosaic interference filter with a 10 nm passband centred on the hydrogen-alpha line; this covers the full 350 mm × 350 mm field of the Schmidt. Impressive fine structure has been revealed in nebulosities in our own Galaxy and in the nearby Magellanic Clouds. These nebulae glow selectively in the hydrogen-alpha region of the optical spectrum and are often so faint that exposure times of up to five hours are necessary to record them (figure 11, plate IV).

Another successful venture has been the construction of a full-aperture 1.2-m diameter narrow-angle objective prism. A single Schmidt prism photograph will record the spectra of several hundred quasars, *about as many as the previously known total number of quasars!* This opens the way to large scale systematic quasar searches, the statistics from which should begin to give significant information on the large scale structure in the most distant parts of the Universe.

One of the most exciting results to come from Siding Spring was obtained with the AAT. A strong radio pulsar with the short period of 89 milliseconds was known to exist in the prominent young (about 10^4 years) supernova remnant in Vela. From theoretical arguments and by analogy with the Crab, this pulsar was thought to be detectable at optical wavelengths. A search was undertaken following a re-determination of the pulsar's radio position and period by a group from Radiophysics CSIRO and the University of Sydney. Optical light from a pulsar is too faint to be registered in a short exposure but must be detected synchronously, adding the results of several hours. If the optical observer searches on a slightly erroneous period any optical pulses will be gradually smeared out. Eight hours' exposure in January 1977, conducted with these precautions, did reveal pulsed optical radiation from a five arc second diameter circle of sky centred on the radio position while a similar exposure on a control area of sky nearby did not. Further observations indicated that the stellar magnitude was 24.2 and established the optimum position.

The light curve (figure 12) proved to be double peaked, with the peaks separated by a quarter of a period. At radio wavelengths the curve is singly peaked and leads the first optical pulse by one quarter of the period. The γ-ray curve is doubly peaked but at a separation of half the period, embracing the optical peaks. This behaviour is in contrast to that of the Crab pulsar whose form of light curve is roughly the same at all wavelengths. The origin of this pulsed radiation is not understood, but it must represent the most concentrated source of light and radio waves so far encountered in the Universe.

Extragalactic astronomy in the United Kingdom has developed rapidly in recent years from the use of a powerful new astronomical detector, the Image Photon Counting System (IPCS), developed by Boksenberg at University College London. This instrument makes it possible to observe in one hour

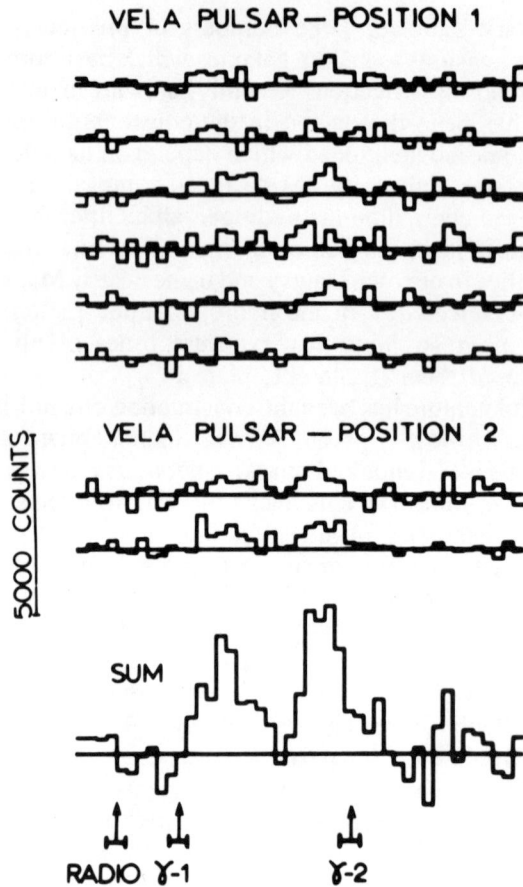

FIGURE 12.　A series of hour-long runs on the field of the Vela pulsar. The pulsed variation is lost in noise on each run but is clearly seen in their sum. The positions of the radio and γ-ray pulses are shown relative to the optical pulses. (*From* Wallace *et al., Nature*, **266**, 692–694, 1977, by permission.)

objects so faint that spectroscopic observations previously required several nights' exposure.

Active galaxies generally appear similar to normal galaxies but have a nucleus of extraordinary luminosity in which light is generated by some non-stellar mechanism giving rise to conspicuous emission lines. A likely mechanism for the enormous energy output of active galaxy nuclei and quasars is the accretion of galactic material on to a central black hole. In support of this, evidence for a black hole in the nucleus of the giant elliptical galaxy M87, which shows many characteristics of an active galaxy, has recently come from IPCS observations of the spatial dependence of stellar motions in this galaxy. There is a sudden rise in stellar velocity dispersion close to the nucleus, which leads to an estimate for the

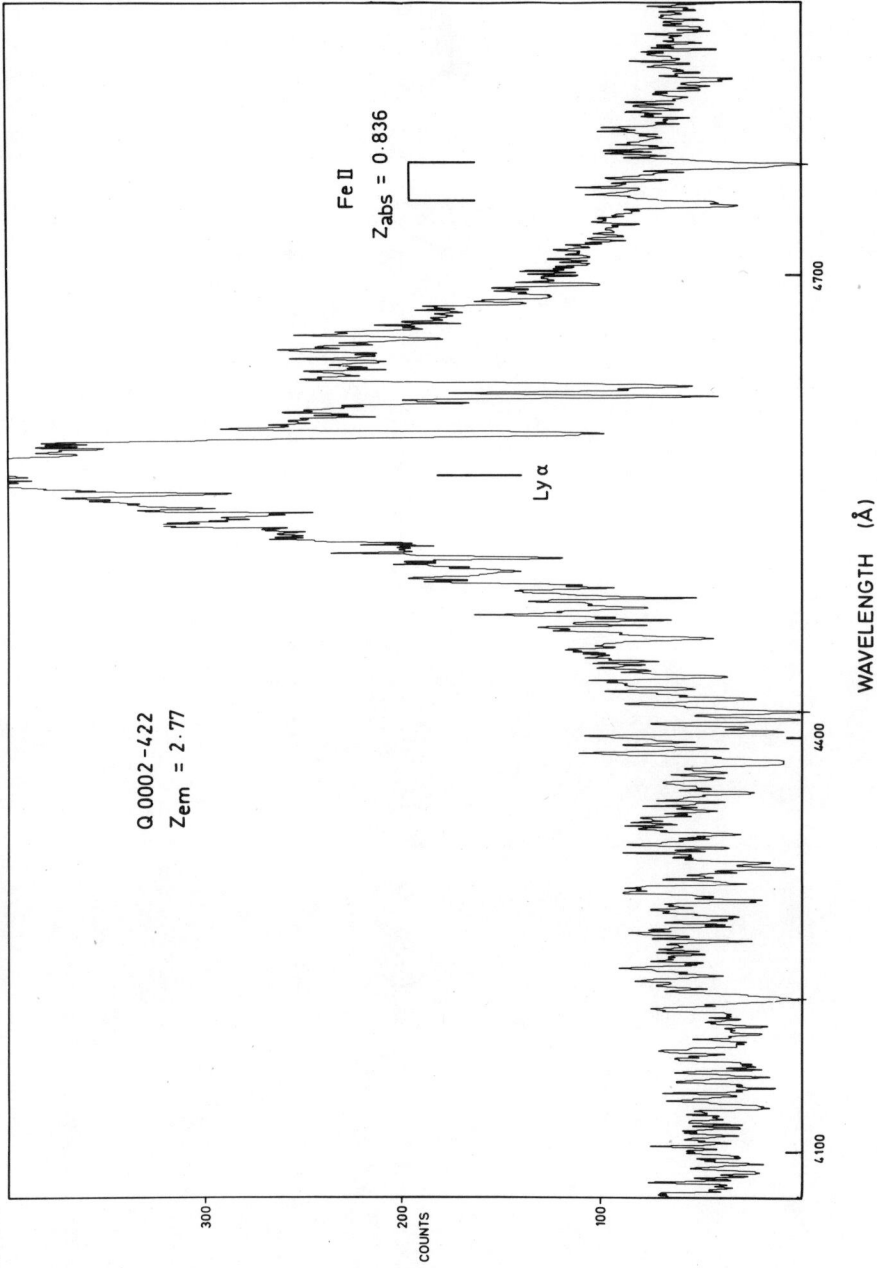

FIGURE 13. Spectrum of the QSO Q0002-422 (obtained on the AAT). Apart from the strong Lyman-α emission line Fe II absorption lines in the system at redshift 0.836 are indicated. This system has a velocity of 0.62 of the velocity of light relative to the QSO and probably is due to interstellar gas in an intervening galaxy.

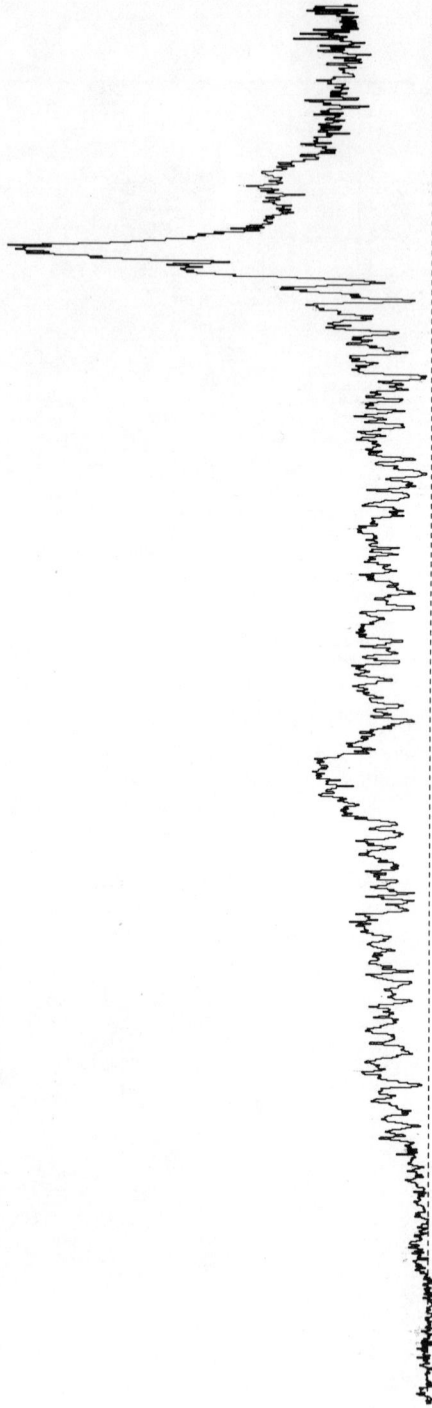

FIGURE 14. Spectrum of the high redshift QSO OH 471. The strong emission feature to the right is the Lyman-α line of hydrogen shifted into the visible region to 5349Å from its rest value at 1214Å in the far ultraviolet, giving a redshift of 3.4.

nuclear mass of 5×10^9 solar masses. However this is not accompanied by a corresponding rise in luminosity which would be expected if this mass were made up of stars. A detailed comparison with model computations shows that the observations are consistent with the presence of a central black hole.

A spectrum of a quasar with absorption in intervening material is shown in figure 13 and another spectrum of one of the most distant quasars so far observed is shown in figure 14. The strong emission feature is the far ultraviolet Lyman-α line of atomic hydrogen which normally is inaccessible to ground-based observatories but here has been Doppler-shifted well into the visible region. This object, with a redshift of 3.4, is receding at close to the speed of light and consequently is near the observable edge of the Universe.

RADIO ASTRONOMY

The past twenty-five years span almost the entire history of radio astronomy. For although extraterrestrial radio waves were first detected in 1932 by Jansky it was the pioneers of wartime radar, Hey, Lovell and Ryle in England, who were to stimulate the rapid development of the subject in the post-war years.

The extraterrestrial radio signals picked up at metre wavelengths by Jansky and later studied by Reber were generated mainly in the interstellar gas. The angular resolution of their aerials was insufficient to discriminate between this and discrete sources. The first discrete radio source Cygnus A was discovered by Hey in 1946, the Crab Nebula was identified as a radio source by Bolton in Australia in 1949 and radiation from the Andromeda Nebula, M31, was detected by Hanbury Brown and Hazard in 1951, while two peculiar galaxies in Virgo and Centaurus had also been tentatively identified by Commonwealth scientists. Although some radio emission is of thermal origin many sources emit by synchrotron radiation from relativistic electrons centripetally accelerated in cosmic magnetic fields.

The discovery of the first radio spectral line, the 21-cm line of atomic neutral hydrogen was made in 1951 by the American astronomers Ewen and Purcell and confirmed by the Australian and Dutch groups in Sydney and Leiden.

Observations on the hydrogen line have had a major effect on astronomy, indeed the importance of this was recognized to be so great that the design of the 250-ft radio telescope then being built at Jodrell Bank was modified during its construction. Originally its surface was to consist of 2-inch wire mesh which would have given it a very low efficiency at a wavelength of 21 cm but it was changed to solid plates with extra support for the weight.

The 21-cm line has been used extensively to study galactic structure and has also proved useful in the measurement of the magnetic field of the galaxy, predicted by the synchrotron nature of the background radiation. Spectral line observations in the United States and at Jodrell Bank in the mid-1960's revealed a small Zeeman splitting corresponding to a galactic magnetic field of about ten microgauss. Doppler shifts in the 21-cm line have made it possible to measure the rotation of galaxies and to determine their hydrogen content. The relationship between mass and galactic morphology may hold valuable clues about galactic evolution.

Some of the most exciting developments in astronomy have stemmed from an increasing awareness of the importance of studying radio waves emitted by objects far beyond the confines of our own galaxy. The earliest catalogues of radio sources had been published by 1952, by Stanley and Slee in Australia (22 sources) and by Ryle, Smith and Elsmore in Cambridge (50 sources). It was immediately evident that the radio sources were not associated with optically bright objects and their true nature only became clear after painstaking efforts to pinpoint their position had been carried out by Smith at Cambridge, and by Bolton and Mills in Australia. High sensitivity photography by the 200-inch optical telescope on Mt Palomar then showed that the prominent radio source, Cygnus A, coincided with a faint optical galaxy having a redshift of 0.057. Cosmological models indicated that this galaxy lay at a distance somewhat greater than 5×10^8 light years and since the radio signal from Cygnus A was relatively strong it was obvious that similar sources could be detected by radio telescopes at far greater distances. Thus radio astronomy offered the possibility of probing the Universe more deeply than ever before.

As always, in a new science, the rate of advance was dictated by technological developments. The need for greater angular resolution, both to investigate the shapes of the radio galaxies and to locate larger numbers of them, led to the construction of radio telescopes radically different from the giant dish at Jodrell Bank. Ryle at Cambridge, and Mills in Australia developed instruments, primarily for sky surveys, employing interferometric techniques, which complemented but did not supplant the dishes. Hanbury Brown, Jennison and Palmer at Manchester sought to resolve the structure of the most powerful sources by supplementing the 250-ft telescope with smaller aerials at distances up to 127 km.

From statistical data, including counts of radio sources as a function of their radio intensity, and the background radiation from the faintest sources (too numerous to be counted individually), Ryle concluded that the distant parts of the Universe were more densely populated with radio galaxies than the nearer regions. This suggested that the Universe must be evolving in time and could not be in the 'steady state' demanded by the Bondi, Gold and Hoyle theory. These studies stimulated much new work at radio observatories all round the world and as we have seen the 'steady-state theory' has been abandoned. It is now generally agreed that the most powerful radio galaxies must be evolving rapidly on a cosmic time scale.

Prior to 1960 it was already known, through the work of Burbidge and others, that radio galaxies posed a substantial problem as to the source of their energy. The discovery of quasars in 1962 made the problem acute. Measurements of the angular sizes of radio sources at Manchester had shown that some of them were exceptionally tiny. One of these, known as 3C 48 (number 48 in the third

Cambridge survey), had been optically identified by Sandage with a blue star-like object. Another, 3C 273, was identified with a similar blue object after its position had been measured very accurately using a lunar occultation. The breakthrough came when it was found that the optical spectrum showed a large redshift. What had previously been thought to be a new kind of star in our own galaxy was actually further off than Cygnus A. When the optical spectrum of 3C 48 was re-examined its redshift distance was found to be more than double that of 3C 273. The radiation from these quasars (quasi-stellar radio sources) showed that they emitted 100 times more light than typical galaxies, that they were amongst the most powerful emitters in the catalogues.

There is no doubt that quasars raise some of the deepest problems in astrophysics. We have noted already the work on them at Siding Spring. No viable alternative to cosmic recession has been proposed to account for quasar redshifts, and it is now generally accepted that they are the most distant objects known to man.

To understand the physical processes that may take place within quasars and radio galaxies it is vitally important to gain further information about their shapes and sizes, which calls for radio telescopes having an extremely high angular resolving power. The concept of aperture synthesis, pioneered by Ryle and his colleagues since 1960 involves the use of an array of small parabolic reflectors, some of which can be moved along rails. By suitably combining the reflectors as an interferometric arrangement it is possible to simulate the behaviour of a giant radio telescope that would be far too large to construct in the normal way. Digital computers are required to process the data and radio telescopes adopting the synthesis technique have advanced hand in hand with the development of such computers.

The One-Mile telescope at Cambridge, completed in 1963, was the first instrument with sufficient angular resolution and sensitivity to provide a clear picture of the structure of radio galaxies. A remarkable feature of these radio maps was that over half of all the powerful radio galaxies showed a characteristic dumbbell shape, in which the radio emission came from a pair of extended clouds lying on either side of a faint optical galaxy or quasar. It seemed that radio galaxies might be explained as some kind of violent explosion in a galaxy whereby the radio-emitting clouds were ejected in opposite directions from a central source of activity.

A more powerful synthesis telescope was completed at Cambridge in 1971. Eight parabolic reflectors were arranged along a line 5 kilometres in length and this provided radio astronomers, for the first time, with radio maps having an angular resolution comparable with that of the best plates that can be taken with a large optical telescope. An exciting new feature that immediately became evident was that the extended clouds appeared to be activated from energetic 'hot-spots' within them. This fact strengthened an earlier theory involving energy beamed outwards from the central source continuously. The hot-spots could be explained as compact regions in which the energy of the beam was conveyed to a host of charged particles, moving at speeds close to the velocity of

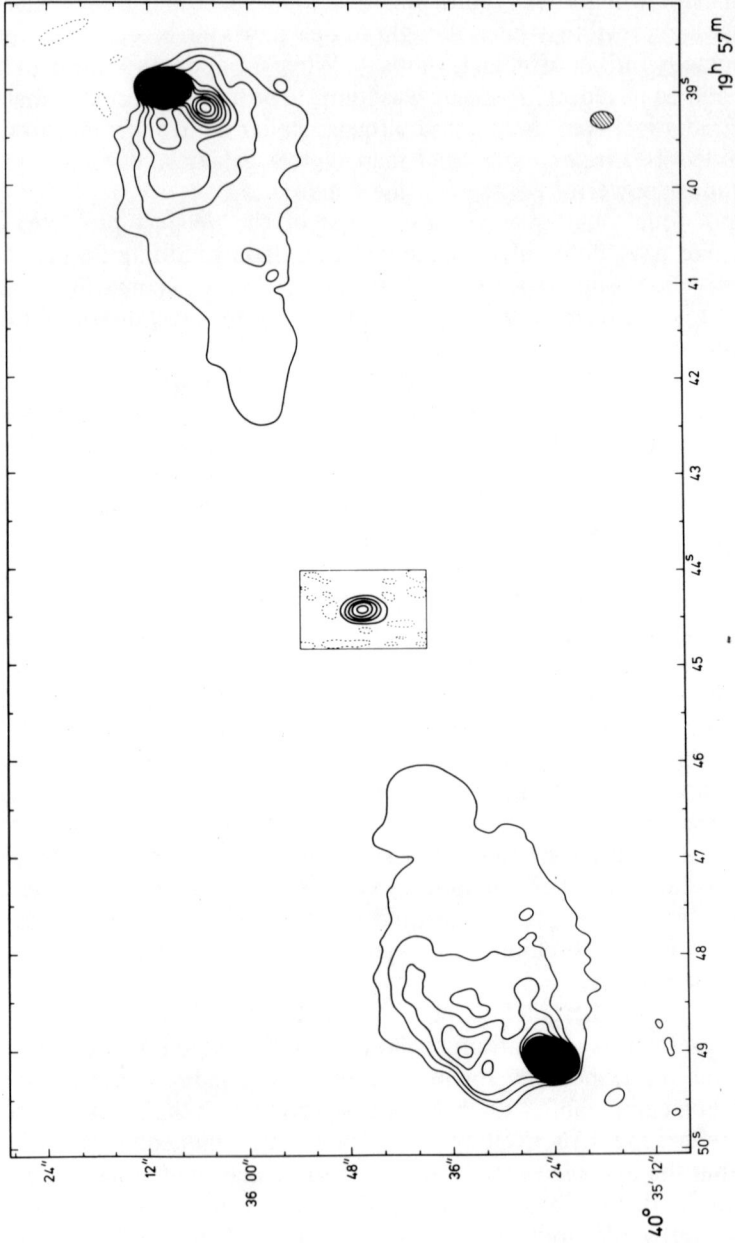

FIGURE 15. A map of the radio galaxy Cygnus A obtained with the 5-km telescope. The contours of radio intensity reveal two compact regions at the outer edges of the larger clouds. These 'hot-spots' are probably energized by narrow beams emitted from the central component which coincides with an optical galaxy.

light, and subsequently giving rise to radio emission via the synchrotron process as they travelled outwards through the clouds (see figure 15).

While quasars were initially discovered through their association with very compact radio sources, the systematic mapping of a large number of radio galaxies, associated with observations using optical telescopes, has revealed quasi-stellar objects close to the centre of many powerful radio galaxies. Optical and X-ray astronomy have also shown that many more nearby galaxies show evidence of violent activity at their centres. The overall picture that emerges is that many galaxies contain an eruptive nucleus and that quasars are the most violent examples of this class.

The nature of these active nuclei is of fundamental importance and we have mentioned evidence that M 87 may contain a black hole. If this is generally the case the energy of quasars would be provided by gravitational collapse. Attempts have been made to obtain much higher resolution information about the structure of compact nuclei by using interferometers of very large baseline (VLBI), as large as the size of the Earth will permit. The 250-ft dish at Jodrell Bank, and similar reflectors in California and the USSR, all make observations of the same radio source simultaneously and the signals are recorded on magnetic videotape. Time signals from atomic clocks are recorded at each site also. Finally all the magnetic tapes are correlated to produce the same result as if the telescopes were connected by cables.

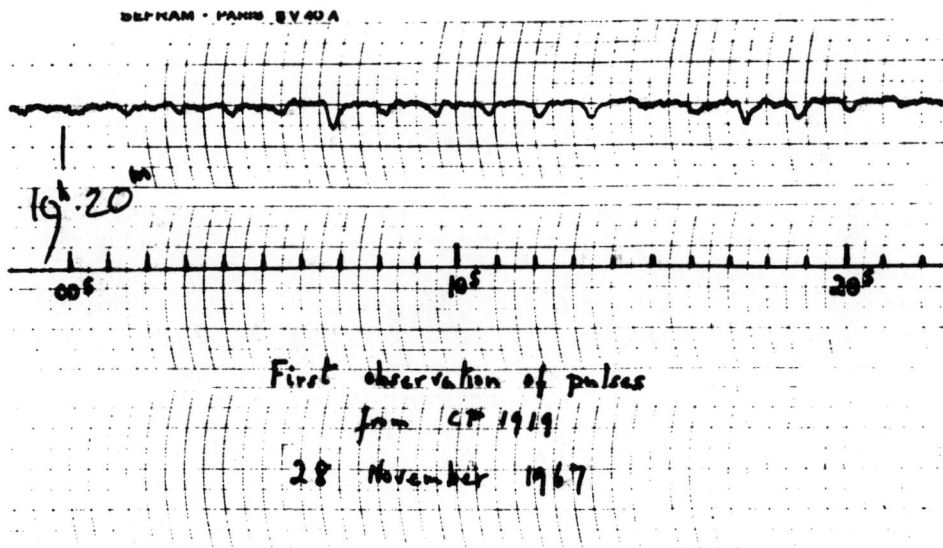

FIGURE 16.   The first observation of a pulsar (28 November 1967); the upper pen shows the pulses received from CP 1919 (intensity increasing downwards!) whilst the lower trace is of time pulses from a clock. The pulsar signal obviously varies markedly in intensity from pulse to pulse, some pulses even being lost in the noise, but the period is accurately maintained at 1.337 seconds. (*Courtesy* Mullard Radio Astronomy Observatory.)

The progress of science is punctuated from time to time by discoveries which were unplanned and unsought. Such was the discovery of pulsars at Cambridge by Bell and Hewish in 1967 as a development of work initiated by Hewish and his colleagues in 1964 when it was found that the radio signals from distant quasars showed a twinkling effect analogous to the scintillation of visible stars. This scintillation is caused by irregular refraction of radio waves by plasma clouds expelled by the Sun. In 1967 Hewish brought into operation a radio telescope specially designed for scintillation observations using a long wavelength particularly sensitive to rapidly flickering signals. Miss Bell (now Dr. Jocelyn Burnell), studying the records of the first survey, noticed a radio source that was remarkable in two ways. It was not always present when it should have been and it scintillated more strongly than usual. A special investigation of this source was carried out using a recorder of higher time resolution and it was found (see figure 16) that the signals consisted of sharp pulses repeated at extremely regular intervals of about 1.3 seconds. Further careful investigation removed the possibility of a terrestrial origin and showed that the emitter could not be larger than a small planet and that it lay far beyond the solar system amongst the nearest stars in our galaxy.

Since 1967 more than 150 pulsars have been located, particularly fruitful surveys being undertaken at Jodrell Bank and at Molonglo in Australia. The radio pulses are emitted by rapidly rotating neutron stars resulting as we have seen from the gravitational collapse of massive stars during the sequence of events associated with supernova explosions. Thus was established the reality of neutron stars, which, prior to the pulsar discovery, had only existed as theoretical speculation.

ACKNOWLEDGMENTS

I gladly recognize the help of Professor Hawking, FRS, Professor Hewish, FRS, Sir Bernard Lovell, FRS, Professor Graham Smith, FRS and of their colleagues who supplied me with much written material. I thank similarly my own colleagues, especially Dr. Bell Burnell who not only assisted with the draft but was also the principal organizer of the Soirée exhibit. Finally I acknowledge that much good work and many discoveries have been arbitrarily omitted. Examples are the Hanbury Brown and Twiss optical interferometry and the British infrared astronomical work and many names that merit a mention in a record like this but for sheer lack of space have had to be left out.

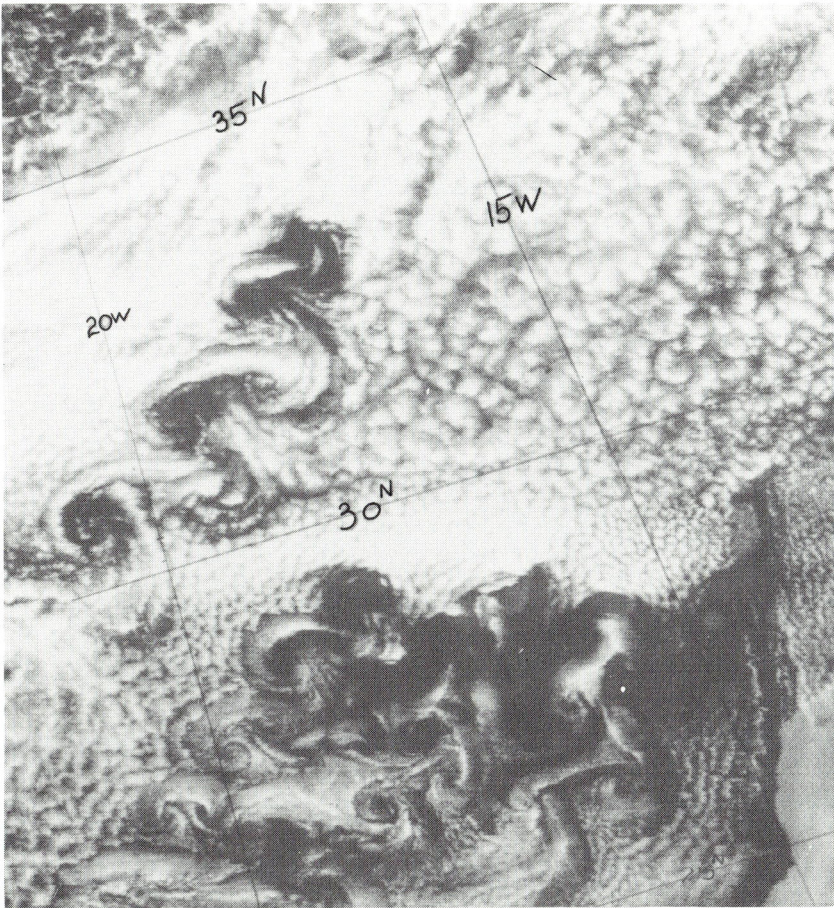

PLATE II (MASON)

FIGURE 2.   A remarkable very high resolution photograph taken by the US NOAA 5 polar-orbiting satellite at 1030Z on 29 April 1977. This shows vortices induced in a steady light north-north-easterly airstream by the Canary Islands and the island of Madeira made visible by the presence of a shallow layer of cumulus and stratocumulus cloud, with base about 2000 ft, lying below the tradewind inversion which had its base at about 3000 ft.

The trains of down-wind vortices are caused by the air flowing over and around the steep-sided mountainous islands which extend far above the temperature inversion. There is a cloud-free area to the lee of the islands where the air is descending but, as the eddies drift downstream, cloud re-forms and the eddies show up as swirls and curved bands of cloud.

PLATE III (MASON)

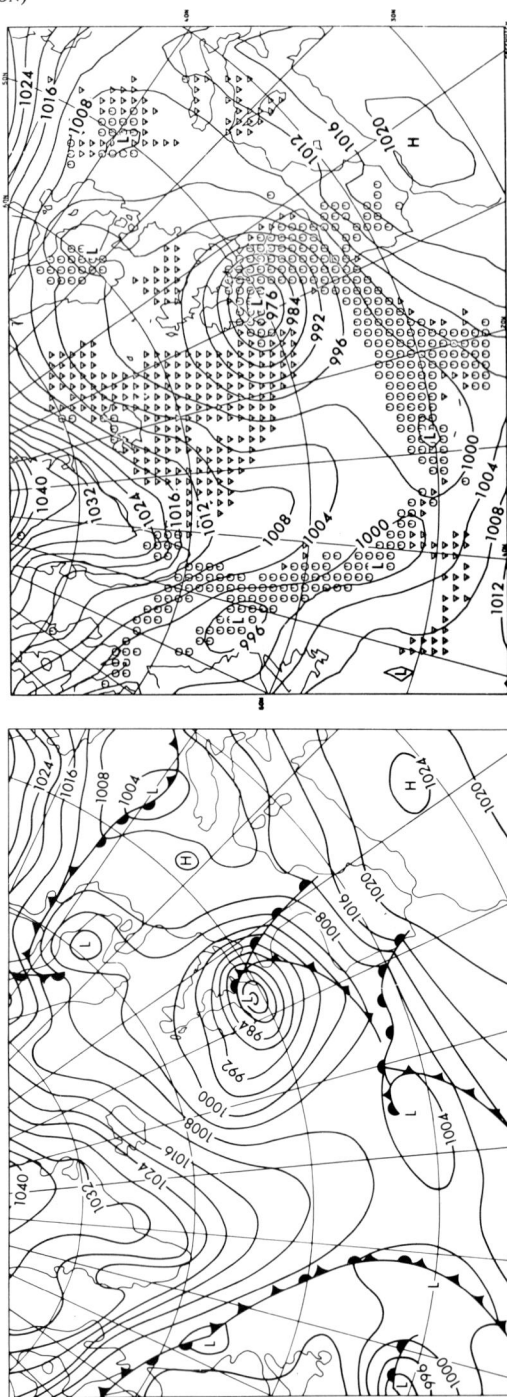

FIGURE 4 (a). The actual distribution of mean sea-level pressure at 12 GMT 12 January 1976.

(b). The corresponding 24-hr numerical forecast of surface pressure and rainfall from the fine-mesh model made at 12 GMT on 11 January 1976.

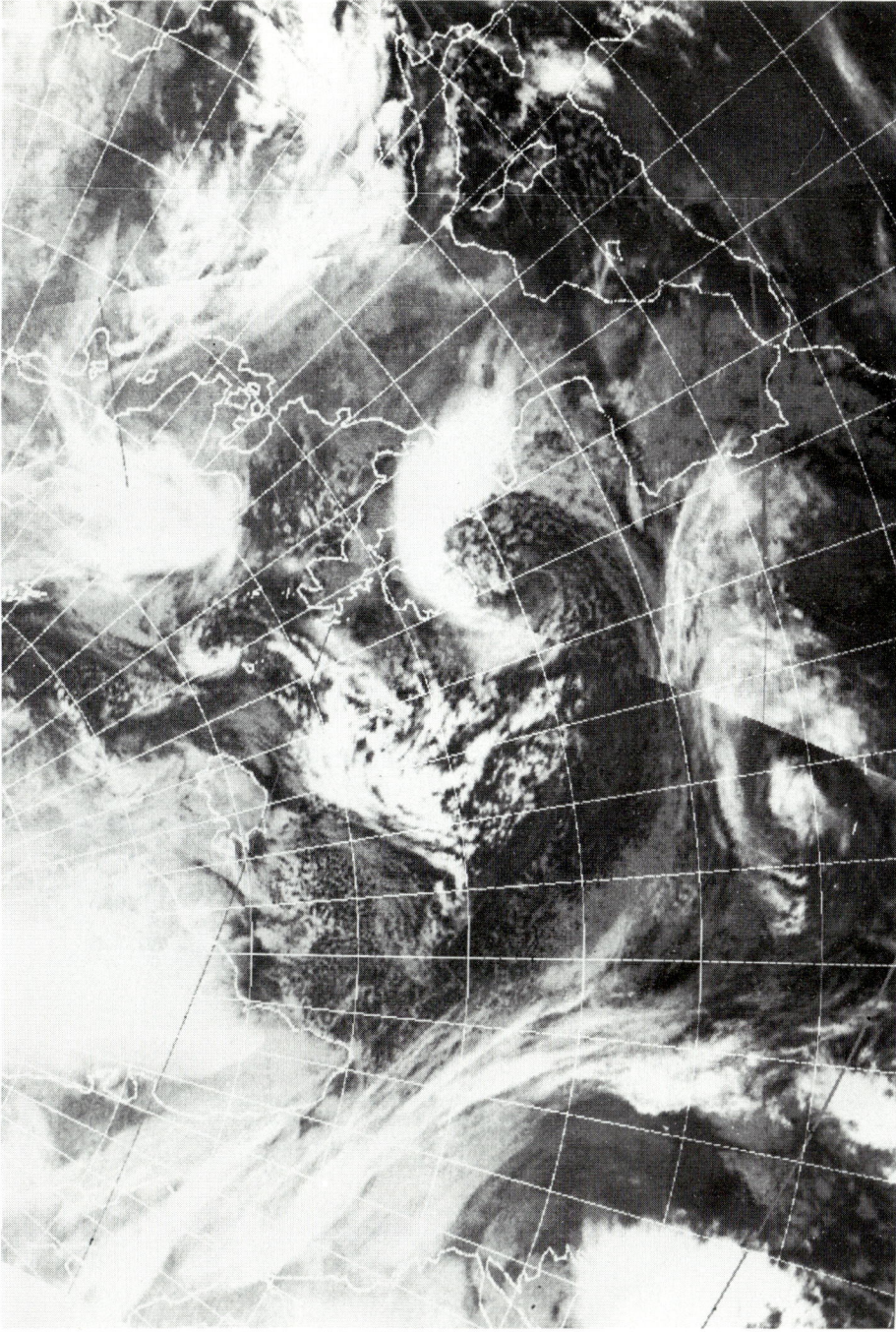

FIGURE 4 (c). Infrared satellite photograph taken at 1153 GMT 13 January 1976 from the US NOAA IV polar orbiting satellite showing good agreement between the actual and forecast positions of the front over southern England.

FIGURE 1. Picture transmitted from the US ATS-3 geosynchronous satellite over the Amazon, showing the whole of South America, North America and Western North Africa. Prominent among the many cloud systems are the spiral pattern associated with a small depression off the coast of Morocco, a long line of convective clouds stretching east–west over the southern part of the North Atlantic, and a cold frontal system over Uruguay and North Argentina. (Photograph by courtesy of US National Aeronautics and Space Administration.)

RECENT ADVANCES IN WEATHER FORECASTING

B.J. MASON, F.R.S.
Meteorological Office, Bracknell, Berks.

INTRODUCTION

During the last decade the traditional, empirical and largely subjective methods of weather forecasting that depend heavily on the experience, skill and judgment of the individual human forecaster, have gradually given way to objective mathematical predictions made with the help of powerful electronic digital computers. The whole operation consists of three stages: data acquisition and processing, analysis, and prediction, and consists basically of forming a three-dimensional representation of the conditions prevailing through a large volume of the atmosphere at a particular moment of time, and of predicting, from an observed initial stage, the future evolution and movement of atmospheric disturbances and their associated weather. Using standardized observations and measurements, made simultaneously at fixed times over a large part of a continent or even a hemisphere, and exchanged rapidly between different countries in universally agreed codes over a special global network of satellite, cable, radio- and picture-transmission channels, charts are constructed to depict the current distribution of atmospheric pressure, temperature, humidity, winds, etc. at the Earth's surface and at a number of levels in the upper air, and from these are evolved forecast charts showing the conditions expected some hours or days later.

WEATHER OBSERVATIONS

In the northern hemisphere, thousands of ground stations and hundreds of ships make observations of temperature, pressure, humidity, wind, cloud cover, sunshine, visibility, rainfall, etc. at regular intervals of one to six hours, and about 600 stations send up balloon-borne radiosondes that telemeter to the ground measurements of pressure, temperature and humidity in the upper air twice daily and radar measurements of winds up to 100 000 ft or more four times daily. These observations are the starting point for the weather analysis and forecast, the quality of which depends a good deal on the coverage and accuracy of these input data. The density of the network is adequate over Europe and

25

North America, but there are large deficiencies over the oceanic and tropical regions and in most of the Southern Hemisphere. Moreover, it is neither economically nor logistically feasible to fill the gaps by extension of existing types of facility such as radiosonde stations, and we must therefore look to satellites to provide global measurements of meteorological parameters by measuring the intensity, polarization, angular and spectral variation of radiation emitted and reflected by the Earth and the atmosphere. In polar orbit, satellites have the great advantage of being able to survey the whole globe every 12 hours and, if placed in sun-synchronous orbit, stay at a set local time with respect to the Earth and so greatly facilitate the ground system for receiving and using the data on a regular schedule. A geosynchronous satellite, i.e. one in equatorial orbit synchronized in period with the rotation of the Earth, remains stationary over a particular point on the Equator, and therefore allows a particular area to be viewed continuously. Current US satellites of this type carry cameras which scan laterally while the satellite spins about its axis, and so cover about one-quarter of the Earth's surface every 20 minutes. They therefore provide a time-lapse film showing the evolution and movement of cloud systems from which valuable information on the winds at cloud level may be deduced. The most recent US satellites provide pictures with a horizontal resolution of 1 km in the visible and 7 km in the infrared from a height of 36 000 km (see figure 1, plate I). A rather similar satellite has been built by the European Space Agency and was launched late in 1977; this is one of an international set of five vehicles built by the United States, Europe, USSR and Japan to provide global coverage during the First Global Observing Experiment in 1978/79.

The latest US polar-orbiting satellites carry scanning radiometers to produce very high quality pictures with a resolution of about 0.9 km in both the visible and infrared parts of the spectrum from a height of about 1000 km and provide day and night coverage (see figure 2, plate II).

Detailed analysis and interpretation of satellite pictures have provided valuable information on the location, structure and evolution of large cloud systems and on such meso-scale features as hurricanes, thunderstorms, squall lines, jet streams and mountain waves, often from oceanic and other remote areas where the observing networks are too sparse to make their detection likely by conventional methods.

Although the cloud pictures and radiation data from relatively simple radiometers currently provide a great deal of valuable information, it is mainly only qualitative in character and difficult to incorporate directly into computer models which require data on atmospheric temperature, pressure, composition and winds. We shall now examine the possibilities of obtaining such data on a global basis by remote sensing of the atmosphere from satellites.

Remote Sensing of Atmospheric Temperature and Humidity

A number of very sophisticated and ingenious spectrometers and radiometers have been developed in the USA and UK and are flying on US satellites to obtain the vertical distribution of atmospheric temperature from measurements of the spectral distribution of radiation received from constant constituents such as carbon dioxide and molecular oxygen.

The radiation emitted by a layer of uniformly mixed gas in local thermodynamic equilibrium is directly dependent on its temperature. But the intensity of the infrared radiation reaching a satellite from, say, the carbon dioxide distributed throughout the depth of the atmosphere, even through a window in which absorption by other constituents is negligible, will also be affected by absorption within the gas itself. Thus relatively little radiation is received from the comparatively small mass of emitting gas in the uppermost layers, and comparatively little also from the lowest layers whose emission is strongly attenuated in traversing the whole depth of the atmosphere. The received radiation therefore comes mainly from intermediate levels and is weighted according to the total atmospheric pressure in such a way that, in effect, the received radiation is a measure of the mean temperature of a layer of atmosphere whose height and thickness are determined by the weighting function.

Moreover, since the absorption is strongly dependent on wavelength, most of the radiation transmitted in the strongly absorbing wavelengths will reach the satellite, while, for less attenuating wavelengths, the received radiation will come from progressively lower levels. Thus a set of measurements of the radiation received at different wavelengths contains information on the vertical distribution of temperature in the atmosphere. In effect, the measured radiances give information on the mean temperatures of a number of overlapping atmospheric layers but it is not possible to derive a unique temperature profile as a continuous function of height from the radiance measurements alone. This requires additional *a priori* information on, for example, the general shape of the profile and the temperatures at singular levels such as the surface and the tropopause either from nearby radiosonde or rocket ascents or from the predictions of a numerical model. Starting with a good 'first guess' temperature profile, the satellite measurements can be used to produce better, up-dated profiles for adjacent times and places and so help fill the gaps between rather sparse conventional soundings. In the absence of such soundings to produce 'first guess' profiles or sufficient additional information to allow a profile to be retrieved from the radiance measurements, the latter can nevertheless provide valuable data on the mean temperatures of atmospheric layers, the heights and thicknesses of which are largely determined by the weighting functions assigned to the various spectrometer channels.

The accuracy of derived temperature profiles is also limited by contamination of the spectral channels by water vapour and cloud. In principle, the cloud problem can be overcome by using microwaves since all but heavily raining clouds are practically transparent to radiation at these wavelengths. Oxygen, which is uniformly distributed up to high levels in the atmosphere, has suitably strong emission/absorption lines at about 5 mm wavelength and these may be used for temperature sounding. An experimental instrument currently operating on a US satellite is producing temperature profiles of quality comparable with those derived from the infrared instrument, and profiles constructed from data provided by both instruments are of higher quality than those based on either instrument separately.

The technique for obtaining temperature profiles from measurements of radiation emitted by CO_2 and oxygen (whose concentrations are known) can also be used to obtain the (unknown) vertical distribution of water vapour if the temperature profile is known and corrections are made for the radiation received from the underlying cloud or land surfaces. Instruments making measurements in the six wavebands of the far infrared spectrum of water vapour are now being used to produce useful humidity profiles in the troposphere.

The Measurement of Atmospheric Pressure and Winds

Accurate determination of the vertical temperature distribution would allow the profile of atmospheric pressure to be calculated provided that the pressure at the Earth's surface is known. Accurate measurement of the surface pressure on land is readily achieved by modern automatically recording aneroid barometers. The problem is much more difficult over the oceans but it is feasible to mount aneroid barometers and other instruments on floating buoys that can be located and interrogated by satellites. About 300 such buoys are being constructed for deployment in the southern oceans during the First Global Observing Experiment.

Determination of the pressure and temperature fields as just described would allow the winds to be calculated in middle and high latitudes where the winds are closely related to the pressure field through the controlling action of the Coriolis force due to the Earth's rotation. This would not be the case in the tropics where the horizontal gradients of pressure and temperature, and the Coriolis force are all weak, so that determination of winds in the tropics remains an important problem. Much effort is now being devoted to the extraction of winds from the movements of clouds on the pictures transmitted on successive frames from the spin-scan cameras carried on geostationary satellites, but this technique can provide only a limited horizontal coverage of winds and information at only two or three levels. In order to obtain additional information on winds over the tropical oceans during the Global Experiment, up to 50 special ships will be equipped with radiosondes, and sondes will also be dropped from aircraft and tracked and interrogated during their fall by equipment aboard the aircraft.

OBJECTIVE NUMERICAL WEATHER PREDICTION

The numerical predictions, with which we shall be mainly concerned in this article, are objective, logical, mathematical exercises based on a firm structure of physical theory that treats the atmosphere as a vast, rotating, compressible, viscid fluid with energy sources and sinks. They involve the construction of physico-mathematical models which, although necessarily rather crude compared with the complexity of the real atmosphere, must nevertheless adequately represent the physical and dynamical processes that are likely to control developments on the space and time scales of interest. In other words, the models must properly represent the relevant or significant scales of motion and their nonlinear interactions, but smooth out all the smaller scale motions that cannot be adequately observed or represented individually while allowing for their overall contribution to transport and energy-conversion processes by representing their statistically averaged properties in terms of larger scale parameters that can be measured. The theory is based on the physical principles governing changes of momentum, mass and energy, on the Newtonian (Navier-Stokes) equations of motion applied to a parcel of air, the laws of thermodynamics, and the equation of state of a gas.

If we consider, at first, a dry atmosphere containing no water substance, the full set of governing equations is as follows:

Equations (1) and (2). Two equations describing the horizontal motions of the air in which the time rates of change of the E–W and N–S components of the wind are related to the forces exerted on the air by the rotation of the Earth, by horizontal pressure gradients, and by retarding forces such as friction and turbulent dissipation of energy.

Equation (3). A similar equation describing the vertical motion of the air under the influence of forces that arise from gravity, vertical pressure gradients, rotation of the earth, and from frictional and turbulent stresses.

Equation (4). An equation of continuity which relates changes in the density and velocity of the air in such a way that mass is everywhere conserved.

Equation (5). A thermodynamic equation which relates the supply of heat to a parcel of air to the resultant changes of temperature and pressure.

Equation (6). An equation of state connecting the pressure, density and temperature of the air.

This set of six equations involves six dependent variables: the three components of the wind, the pressure, density and temperature of the air, all expressed as functions of space and time. In order to include the effects of evaporation, condensation and precipitation of moisture, equations are added for the continuity of the water substance and the heating term is modified to include the release of latent heat.

Starting from a given initial situation and specified boundary conditions, the problem is to solve a system of simultaneous nonlinear partial differential equations in three spatial dimensions with time as the fourth independent variable. In fact, the equations are formulated in such a manner that they allow the time variations of the above quantities to be determined from their spatial variations. Thus, in principle, if we can observe the initial values of all the variables at a network of discrete points filling the whole or a large part of the atmosphere, we can compute the initial time rate of change of each variable from the governing equations, and then extrapolate over a short time interval to find a *predicted value* at each point in the network. Repeating this process step by step, we can build up a forecast of the fields of pressure, wind, temperature, humidity, etc.

In practice, the meteorologist measures the horizontal winds (by tracking balloons with radar), the atmospheric pressure, temperature and humidity, but not the vertical component of the air motion since this is usually too small to measure directly although of the greatest importance, particularly in controlling cloud and rain forming processes. In order to eliminate the density of the air as a variable from the equations, we use pressure rather than height as the vertical coordinate, regarding this as an independent variable and introduce, as dependent variable, the contour height, which is the height of a constant pressure (isobaric) surface. Maps of contour heights are very similar, in configuration and in their relation to the winds, to those of the corresponding pressure fields. The thermal structure of the atmosphere is described in terms of the thickness of the layers between isobaric surfaces, the thickness being proportional to the average temperature of the layer. In pressure coordinates, the vertical velocity of the air is represented by the time rate of change of atmospheric pressure, and is defined entirely by the horizontal wind field. Further simplification is introduced by rewriting the thermodynamic equation in a form that relates the vertical motion to changes in the non-adiabatic heating and the thickness of the layer, so that the dependent variables become the two horizontal components of the wind, the time rate of change of the atmospheric pressure (closely related to the vertical motion), the contour height, thickness, and humidity of the layers between isobaric surfaces, from which quantities such as the precipitated water (rain and snow) can be deduced.

Such complex physico-mathematical models have been developed in the Meteorological Office for use on its powerful IBM 360/195 computer to replace a simpler model which formed the basis of its daily forecasting operations between 1965 and 1972.

The current model is designed to simulate and predict the evolution of major weather systems over practically the whole of the Northern Hemisphere for several days ahead. The basic governing equations are essentially those described earlier and include those for the continuity of moisture. If a layer of air is unsaturated, the humidity is allowed to change by horizontal and vertical air motions and by evaporation of rain or snow falling into it from above. Once a layer reaches saturation, its excess moisture is deemed to condense and fall out as rain or snow (depending upon the temperature) into the layer below. The latent heat released or absorbed during the processes of condensation, evaporation, freezing and melting is calculated and its dynamical consequences computed. The model allows for modification of the airflow by the underlying topography, for the frictional drag of the land and sea on the air and for horizontal eddy diffusion. Adjustments are also made to allow for the vertical transport of heat and moisture by both shallow and deep convective clouds that are smaller than the computational grid. In computing the exchanges of sensible and latent heat at the Earth's surface, the surface albedo is specified at each grid point in terms of the climatic conditions and the snow or ice cover, and sea-surface temperatures are held at their monthly mean values. Even for short-period forecasts of 2–3 days, it is necessary to compute the heat lost by the atmosphere through long-wave radiation to space and the long-wave transfer between model layers including the effects of clouds and the ground, otherwise predicted temperatures for the middle and upper levels are too high.

Numerical integration of the equations was at first carried out using a straightforward explicit Lax–Wendroff finite-difference scheme, but this has a severe stability criterion set largely by fast-moving gravity waves which are of little meteorological significance but permit a maximum time step in our model of only 400 a ($6\frac{2}{3}$ min). A forecast for 72 hours ahead then involved about 2×10^{10} numerical operations and took about 42 minutes on the IBM 360/195 computer operating at about 10 million instructions per second (10 Mips) on this type of problem. This scheme has now been replaced by a split explicit scheme with the governing equations split into two stages. The changes due to the acceleration and divergence terms that are largely responsible for the gravity waves are computed in time steps of 10 minutes and separately from the nonlinear advection, frictional and diffusion terms which are integrated over 30-minute time steps. This new scheme together with some other minor economies has reduced the computing time by a factor of 6, the integrations for a 3-day forecast now being completed in only 7 minutes.

In addition to this hemispheric model, the Meteorological Office runs a finer-mesh model, with very similar dynamics and physics but with a horizontal grid length of only 100 km limited to a rectangle 6400 km × 4800 km centred on the British Isles. This provides more detailed forecasts for the UK and Western Europe up to 36 hours ahead and is particularly useful in making quantitative predictions of rainfall from depressions and fronts and distinguishes between widespread persistent rain and showery convective rainfall. Since the time steps of the integration are one-third of those for the hemispheric model, the amount of computation is about the same.

Both models are run twice a day using the noon and midnight observations according to the following operational sequence. When the weather observations from practically the whole of the Northern Hemisphere, amounting to about $1\frac{1}{2}$ million coded groups per day, arrive at Bracknell, they are automatically checked, quality-controlled, corrected, edited and arranged in suitable format for input to the models by two dedicated smaller computers. The next stage is to produce analyses of the basic data and from these establish initial values of the dependent variables in the forecast equations. This involves the production, by various objective smoothing, curve- and plane-fitting techniques, of smooth continuous fields of contour heights and humidity mixing ratios from which can be extracted grid-point values over the whole area and for the ten levels. The analysed fields are now subjected to an 'initialization' procedure to ensure consistency between the contour and wind fields. This involves the solution of two equations, one dealing with the non-divergent component of the wind field and the other with the divergent component and

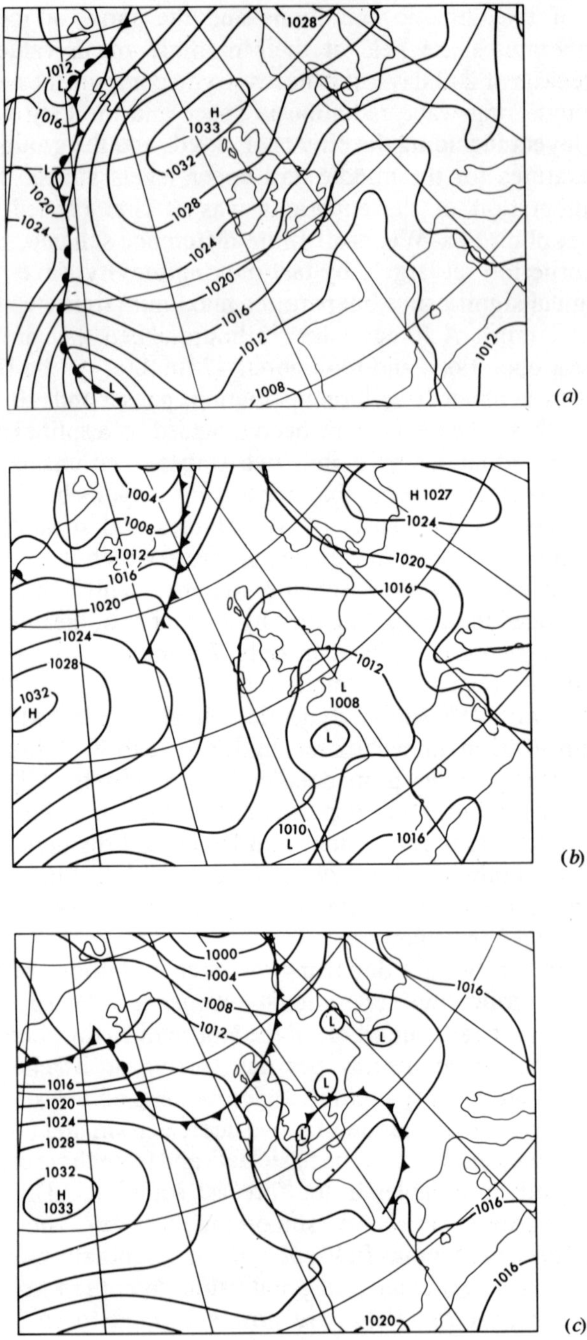

FIGURE 3 (a-c). Maps showing the actual (observed) distribution of surface pressure at 12 GMT on the 28, 29, 30 August 1976.

hence the vertical motion, to produce 'best' initial values of the contour height and the three wind components. These parameters together with grid-point values extracted from a smoothed initial humidity field are stored in column format together with geographical, topographical and climatic constants, surface temperature and humidity parameters, to provide the starting conditions for step-wise integration of the predictive equations.

Some Examples of Numerical Weather Forecasts

Figures 3 (*a–c*) show a sequence of hand drawn analysis charts constructed from observations at 12 GMT on 28, 29, 30 August 1976, illustrating the breakdown of the memorable drought of that year. A so-called 'blocking' situation of extreme duration had developed in which an anticyclone remained centred over or near the United Kingdom for months on end. Towards the end of the drought, the high moved North of Scotland (1*a*) bringing a hot dry easterly airstream over most of the country, with occasional weak troughs of low pressure affecting the South East of England. Breakdown of the drought was effected by the formation of a thundery depression over Northern Spain which moved northwards across Biscay into the English Channel bringing heavy rain to the South of England. At the same time the anticyclone North of Scotland rapidly declined and lost its identity within about 12 hours.

The numerical forecasts from 12 GMT 27 August 1976 (shown in figures 3(*d–f*) drawn by the computer on a cathode-ray tube), correspond closely to the charts of the actual weather in this area, particularly in the predicted sequence of events. The 24-hour forecast shows the gradual emergence of a weak trough of low pressure over Spain which has developed into a separate depression *in situ* by the time of the 48-hour forecast, with a trough extending northwards into Northern France. The anticyclone is forecast to be declining, but retaining its identity, whereas at this time it had in fact almost disappeared. By 72 hours, the forecast shows the northward movement of the depression into North Biscay, while the anticyclone has lost its identity. The timing error of 12–18 hours in the 72-hour forecast is more than compensated by the fact that the model predicted the formation of the low over Spain, and also the major change of weather type at the end of a prolonged period of drought.

In figures 4(*a–c*), plates III, IV, a 24-hour forecast made with the fine-resolution model from 12 GMT 12 January 1977 data is compared with the corresponding actual weather chart and also an infrared satellite picture for approximately the same time, i.e. 12 GMT 13 January 1977. A weak trough in the North Atlantic developed into a small low which explosively deepened and moved very rapidly eastwards into South West England bringing gales and blizzards to much of Southern England. The forecast proved to be correct not only in the position and intensity of the system, but also in the position and nature of the precipitation associated with it. The round symbols in the numerical forecast represent continuous rain or snow falling from the frontal system on the

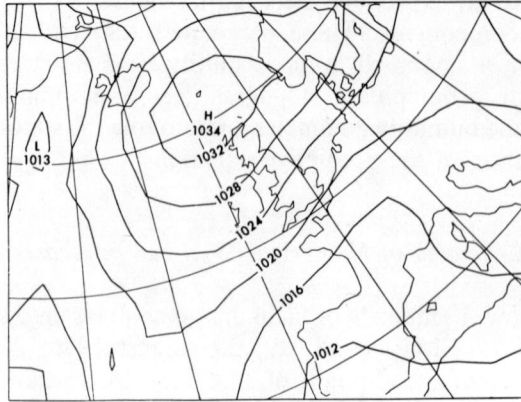

(d)

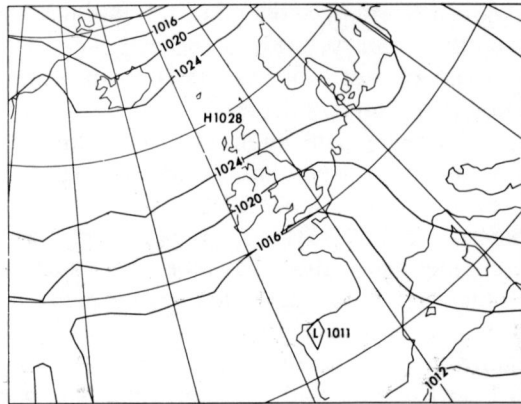

(e)

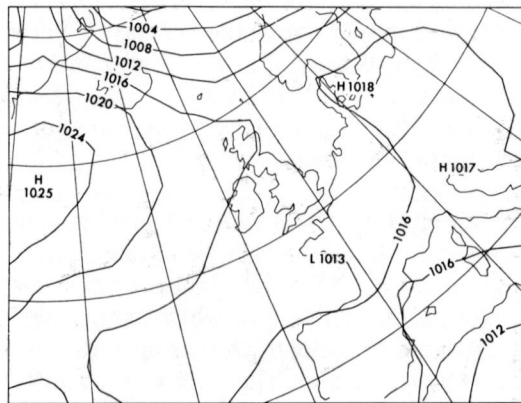

(f)

FIGURE 3 (d-f).   Maps showing the corresponding numerical
forecasts of surface pressure made at 12 GMT on 27 August
1976 for 24, 48 and 72 hrs ahead.

southern side of the depression shown in both the analysis and cloud picture. Forecast temperatures indicated that any precipitation would fall as snow over England. Another feature of interest is the area of thunderstorms to the northwest of the system, shown in the cloud picture. This corresponds to an area of triangular symbols in the forecast representing showery or thundery rain.

Performance, Problems and Prospects

A decade ago, before the introduction of computer models, forecasts were rarely issued for more than 24 hours ahead and these were both less accurate and less detailed than those of today. The numerical methods have led to a greater degree of continuity, consistency and confidence in the forecasts than existed when they depended entirely on the personal judgement of changing rosters of forecasters. The human forecaster still has a vital role to play in interpreting the computer products and in predicting the behaviour of smaller-scale weather systems and local phenomena such as showers, thunderstorms, fog, frost and icing which are not treated by the models, but he depends heavily on the computer forecasts of the broader-scale developments and of marked changes of weather type. The present ten-level models now lead to 72-hour predictions that are as good as 48-hour predictions were a few years ago, and 48-hour forecasts about as good as 24-hour forecasts were then. Moreover the proportion of serious errors in the 24-hour forecasts has been halved from 13% to 6% whilst the proportion of very good forecasts for the British Isles has increased from about 50% to 70%. About 85% of forecasts for the following day for a given region of the British Isles are essentially correct, with about one in six containing significant errors, notably in the timing of the arrival of weather systems from the Atlantic Ocean where observations are sparse.

Although forecasts issued to the general public are usually limited to three days ahead, the hemispheric model is integrated up to six days ahead once a day and the results provide the basis for the weekly farming forecast issued on Sundays. Although the model remains stable for much longer periods, the detailed predictions tend to deteriorate rather sharply beyond the fourth or fifth day. Reliable forecasts for a week or more ahead, which would be of great economic value to weather-sensitive industries such as agriculture, building, fuel and power and food manufacture, will require: improved models, especially in the treatment of clouds and of radiative and turbulent transfer processes; geographical extension to cover the tropical regions down to perhaps at least 20°S; improved observations from the oceanic and tropical regions; and greater computing power. It is important to realise, however, that there are *inherent* limitations in the predictability of atmospheric behaviour, set by the cumulative effect of random small-scale disturbances than cannot be directly observed or treated explicitly in the models but which introduce 'noise' into the system and ultimately destroy the forecast. Nevertheless, current research encourages us to believe that it should be possible to forecast the evolution of major weather systems such as the mobile depressions of middle latitudes for about a week or ten days ahead and to give a useful indication of general trends over periods of perhaps a month. In some recent experiments in which a five-level hemispheric model was run for up to 50 days and the results averaged over ten-day periods, the development of the long planetary waves, which largely determine the general character of the weather over periods of several weeks, was predicted quite well. Also the first attempts to produce forecasts for a month ahead by this method are distinctly encouraging and hold some promise that these objective dynamical methods may ultimately replace the present empirical and largely analogue methods of long-range forecasting.

Although dynamical models have produced substantial improvements in the accuracy of forecasts for 24–72 hours ahead, they have contributed rather little to forecasts of surface weather, especially of the timing, duration and intensity of precipitation, over the first 12 hours. This is partly because the models, in starting *ab initio* from new observations every 12 hours, take some hours to recover from the smoothing and balancing of the initial fields, and also because the weather for the following few hours is often dominated by small-scale systems that cannot be represented explicitly in the current models. Although the Meteorological Office is developing high resolution models with grid lengths of only 10 km in attempts to represent and predict these smaller-scale systems, their enormous computing requirements will probably preclude their operational use for some years. Meanwhile, practical improvements in short-range forecasting may be expected from better radar and satellite observations.

POSTSCRIPT

The staff of the Meteorological Office take particular pride in being in the very forefront of numerical weather forecasting because it started there with the great L.F. Richardson, FRS (1881–1953) who, thirty years ahead of his time, first conceived of predicting the weather from the basic hydrodynamical and thermodynamical equations. He set this out in great detail, including numerical methods for integrating the equations, in a remarkable book partly written in France during the First World War where he served in an Ambulance Unit. The manuscript was lost but later discovered under a heap of coal in France in 1919 and finally published by the Cambridge University Press under the title *Weather Prediction by Numerical Process* in 1922. Richardson's scheme, which is essentially similar to that used today, could not then be implemented because he calculated that he would require 64 000 staff with desk calculators to keep up with the weather. He imagined these assembled like a vast orchestra in a circular arena with a conductor (master programmer) in a central pulpit conducting the whole exercise by directing beams of rosy light on those who were running ahead of the rest and blue light on those falling behind! Even so, he greatly underestimated the magnitude of the task and his heroic attempt to calculate the change in pressure at one point in the Earth's surface failed partly because of computational truncation errors but more seriously because of lack of information on the convergence/divergence of air at higher levels. However, Richardson foresaw and discussed many of the serious problems which we still face. If he were to walk into the Forecasting Centre at Bracknell today he would undoubtedly be at home with the science if a little overawed by the technology.

RESEARCH IN SEAS AND OCEANS

Sir Peter Kent, F.R.S.

Natural Environment Research Council, London

Introduction

The oceans and seas provide recreation, food, mineral wealth, transport, a generator and regulator of world climate and the universal sink to which all land generated detritus and pollution ultimately goes.

From the earliest times man has been concerned about the winds and currents of the sea, about its storms and erosive power, and from the time of the first sailors man has dumped his waste material into it. With increasing pressure on all land-based resources we now have to take much more account of the seas and oceans, as part of the world's energy system and for the means to sustain human life. We have to develop our understanding of the physical, chemical and biological processes involved, how to use them and how to avoid permanent damage through ignorance or inadvertence.

It is fortunate that the sharp increase in the need for scientific data can be supported by the results of decades of research, for in the environmental field the range of variation of natural events—from droughts and tidal surges to earthquakes and insect plagues—can only be understood and analysed from a lengthy build-up of data. But the past twenty-five years have seen major advances in many aspects of natural science, and not least in the marine realm, based partly on the steady accumulation of data, partly on new methods and new concepts.

Some of these advances are outlined in the following pages.

Studies in the Shallow Seas

The estuaries and shallow seas are the areas of most immediate importance to maritime nations, for economic and aesthetic reasons and also as the locus of natural disasters resulting from storms, erosion, wave forces and tidal surges.

In the biological realm fish resources represent the ultimate item in a long food chain, relating back through phytoplankton to the varying physical and chemical characteristics of the sea water, to seasonal changes and longer term variations, both natural and man induced, cyclic and irreversible. Understanding of these

39

inter-related factors has made major strides in recent years as different branches of science have become integrated, but there is still much to learn about the factors which control marine life and resources.

Nature has its own ways of accommodating pollutants. Some of the material discharged into the shallow seas is flushed out into the much larger body of ocean waters, thereby greatly diluted. In estuarine environments we can trace the adsorption of toxic metallic residues on the clay minerals of the muds—rendering them harmless provided quantities are not too large. Estuarine bivalves too can accommodate toxic materials (within limits) by secretion in particular parts of the organism; this material however is released after death, and may poison higher, predatory animals. All these processes depend—like the oxidation of human waste in inland waters—on the system not being overloaded, and our progressive understanding of the mechanisms involved is important in the management and dictation of policy for food resources and waste disposal.

Tidal prediction has a very long history; it is still an essential aid to navigation, becoming increasingly sophisticated as ships become larger and navigational margins more limited. Wave prediction is a newer requirement, arising over the

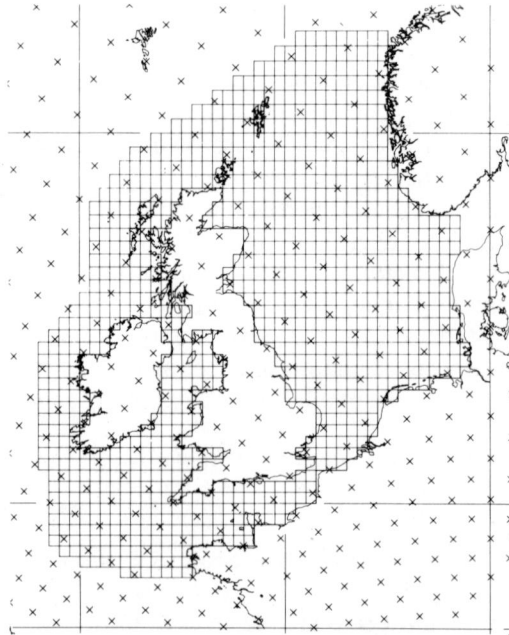

FIGURE 1. A computer model of the effect of wind and pressure on the UK continental shelf allows prediction of water levels and currents to be made up to 30 hours into the future. Forecast wind and pressure values are applied to the model at the points X; when combined with predicted tidal heights at the continental shelf edge the surge heights and current strengths can be computed at intersections of the rectangular lattice.

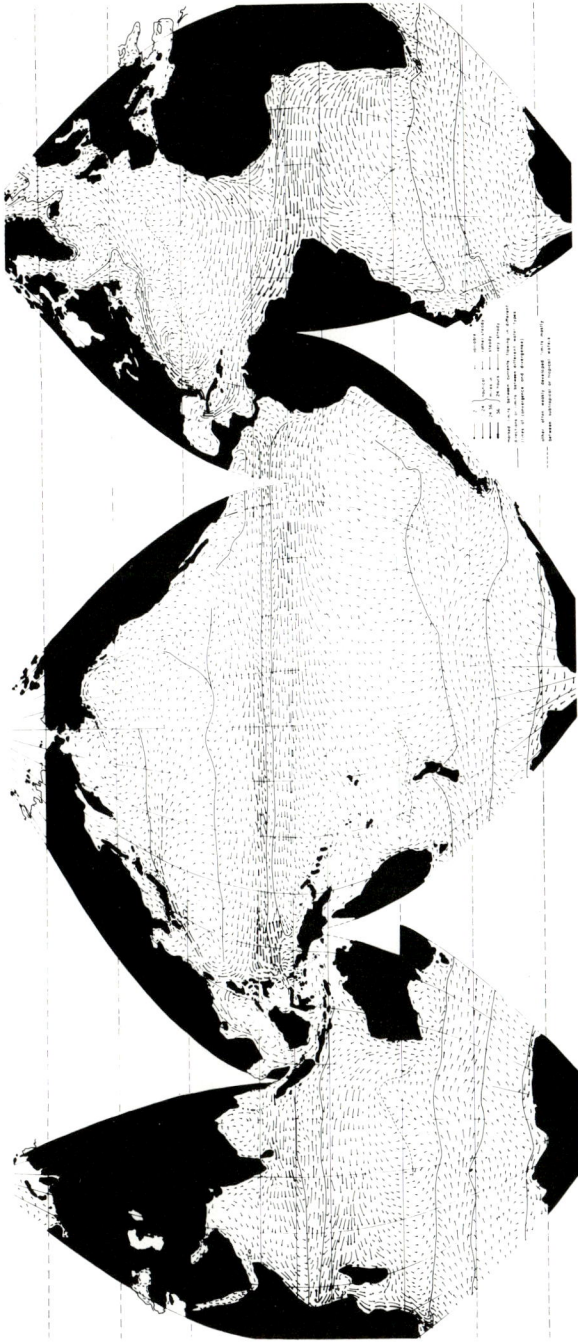

FIGURE 3. Map of surface ocean currents, derived from ship drift observations.

PLATE II (KENT)

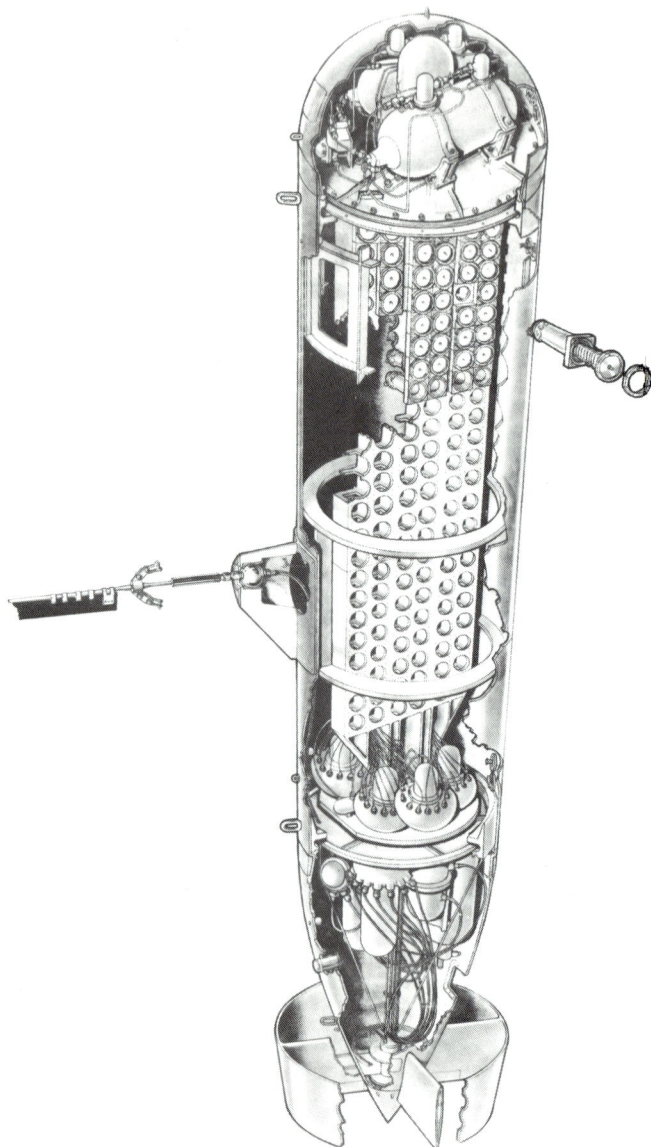

FIGURE 6. An exploded view of the GLORIA Mark I sonar vehicle. The body, 10 metres long, is made of glass fibre reinforced by aluminium bulkheads, and weighs 7000 kg in air. It carries an array of 144 acoustic transducers which may be remotely tilted so that the sound is radiated through the side at the correct angle to the sea floor.

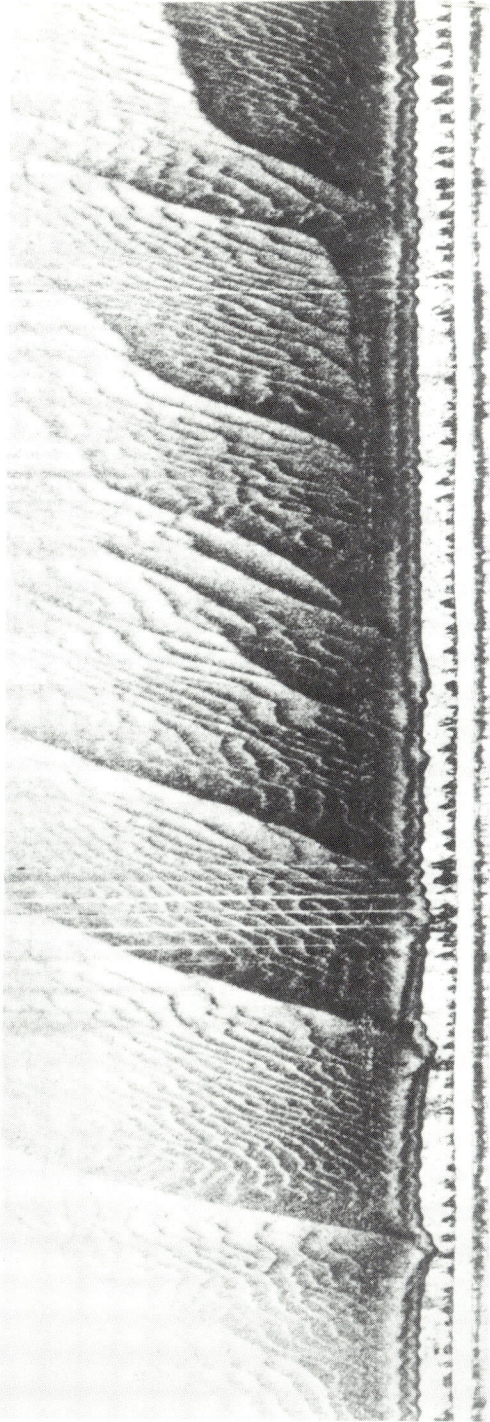

FIGURE 7. Example of a sonograph of sand waves.

Plate IV (Kent)

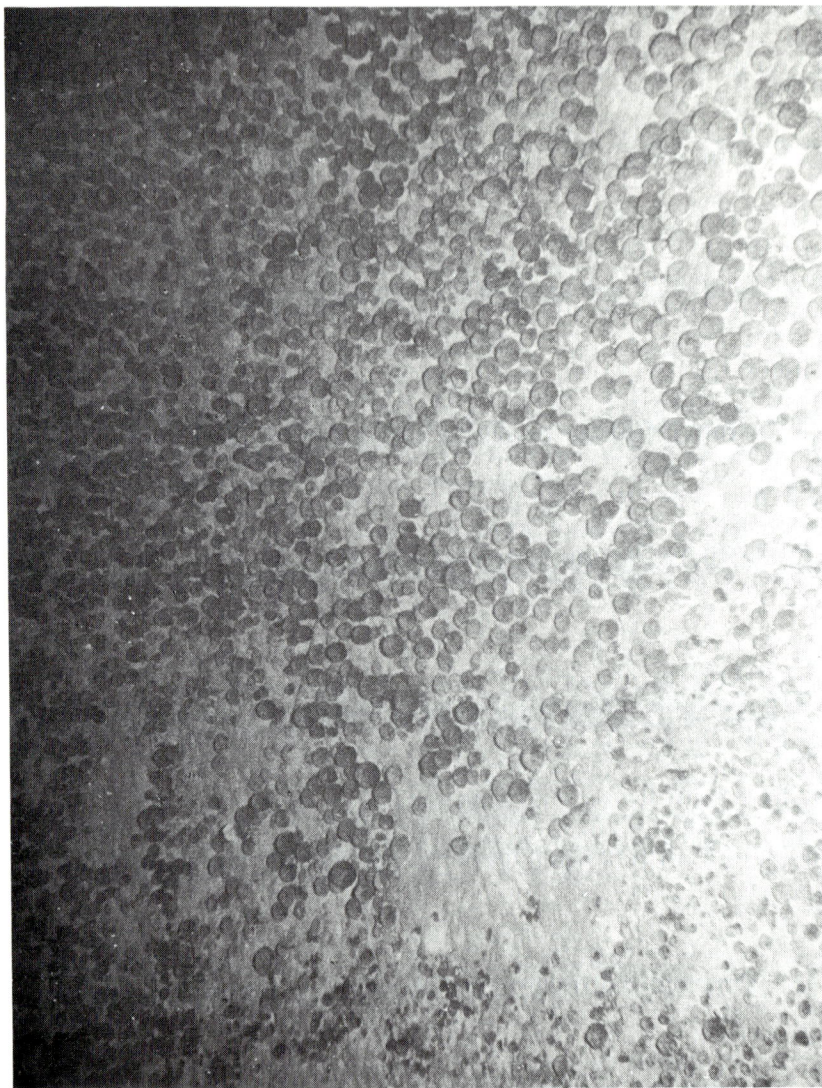

Figure 11. Manganese nodules on the deep ocean floor.

last three decades and now sharply focused on the need to design offshore structures in the North Sea and other areas, where safety considerations must be paramount but a change of 5% in the estimate of maximum wave size relates to millions of pounds difference in cost. The background of decades of data held by the Institute of Oceanographic Sciences was essential for the engineering operations. Because it had successfully developed methods for offshore data collection it was possible to extend the areal coverage with consequent improvement in the predictive ability. Tidal surges are a different problem in this field (figure 1); they are directly relevant to the adequacy of coastal protection, to coastal flooding and the hazard of flood disaster to major cities, including London. Negative surges (abnormally low water) are the complementary problem, bearing directly on the safety of large tankers and bulk carriers, which may have clearances of only a metre below the keel when navigating shallow seas.

The sea floor and its changing topography are the subject of sedimentological studies, of interest on the one hand to the understanding of ancient sediments some of which provide source rocks and traps for hydrocarbons, on the other hand to the problems of maintaining shipping channels, of sources of offshore bulk minerals (sand and gravel, in some places also tin-bearing and hence a source of metal ore), as the foundation for offshore structures and for their bearing on submarine pipeline and cable routes. The sediments are nearly as varied as the range of problems, but we now know much more about their distribution round Britain, and are developing detailed knowledge of the dynamics, for example, of the shifting submarine sand waves, which are a major feature of the Straits of Dover and the Southern North Sea (figure 2).

An unexpected result of detailed side-scan sonar surveys of British sea floors has been the identification of deep scars left by grounding icebergs on the edge of

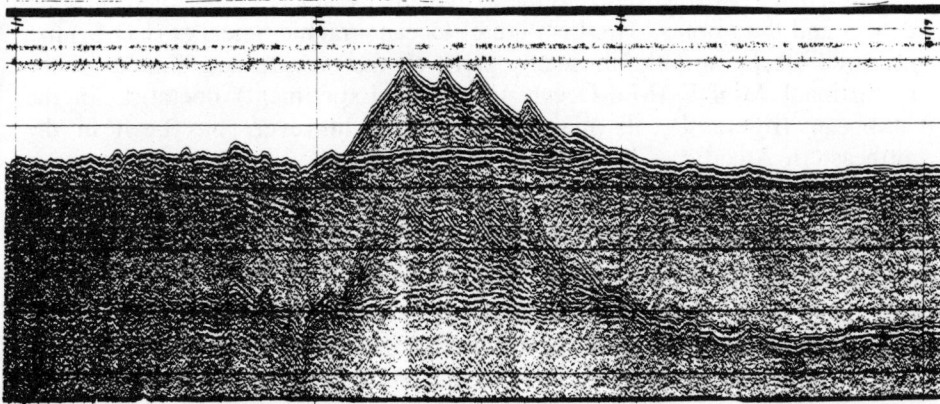

FIGURE 2. Section through mobile sand waves. In areas of mobile sediment such as sand wave fields, pipelines may be re-exposed or even left unsupported.

the continental shelf and upper continental slope made at a time of lowered sea level.

The study of solid rocks below the sea bed on continental shelves has now achieved a new importance with the world-wide search for oil resources to augment and replace the diminishing reserves on land; this subject is described in the final section of this chapter.

Oceanic Circulation

Until the early 1950s our knowledge of the surface circulation of the world's oceans away from energetic and well defined currents such as the Gulf Stream was rudimentary and was derived from ship drift observations (figure 3, plate I). Charts of the currents tended to exaggerate the simple gyre structure of the surface flows. More was known about the surface currents such as the Gulf Stream from observations by mariners of the drift of ships, but the estimates were averages over a period of time and were limited to shipping lanes.

Data from surface ships have now been supplemented by satellite observations, using temperature sensors, which show (for example) that the Gulf Stream has great meanders and swirls.

One major step forward in the studies of the deep circulation was achieved by the British development of neutrally buoyant 'SOFAR' floats, which can be set to drift with the water body at a pre-determined depth, and are tracked acoustically from an attendant ship. These floats have shown that deep currents are very largely different from those on the surface. A further development combined the use of arrays of sub-surface moored recording current meters, which can be left for periods of up to nine months.

Research of this kind has demonstrated tidal oscillations and large vertical eddies in the deep ocean, and (together with deep chemical analyses) has established that oceanic water circulates over great distances—between the polar regions and the tropics. The Institute of Oceanographic Sciences (I.O.S.) has measured deep currents in many parts of the world, and shared in the international MODE (Mid-Ocean Dynamics Experiment) operation in the Caribbean (figure 4); its main contribution, however, has been in the northeastern Atlantic.

Studies in the field of oceanic circulation have many aspects. One of these is military, in relation to the use of submersible vessels, but this lies outside the conventional research field. Another relates to the possibility of disposing of radioactive waste on (or in) the ocean floor, involving the problems of dispersion, dilution and the ultimate fate of such material. In terms of the life of radioactive elements this concerns not only the present currents down to the ocean floor, but also the changes in the current system over hundreds of thousands of years, in glacial and interglacial periods.

Still another aspect of oceanic circulation and its variability follows from the close interaction of ocean water and the atmosphere, and the major control of

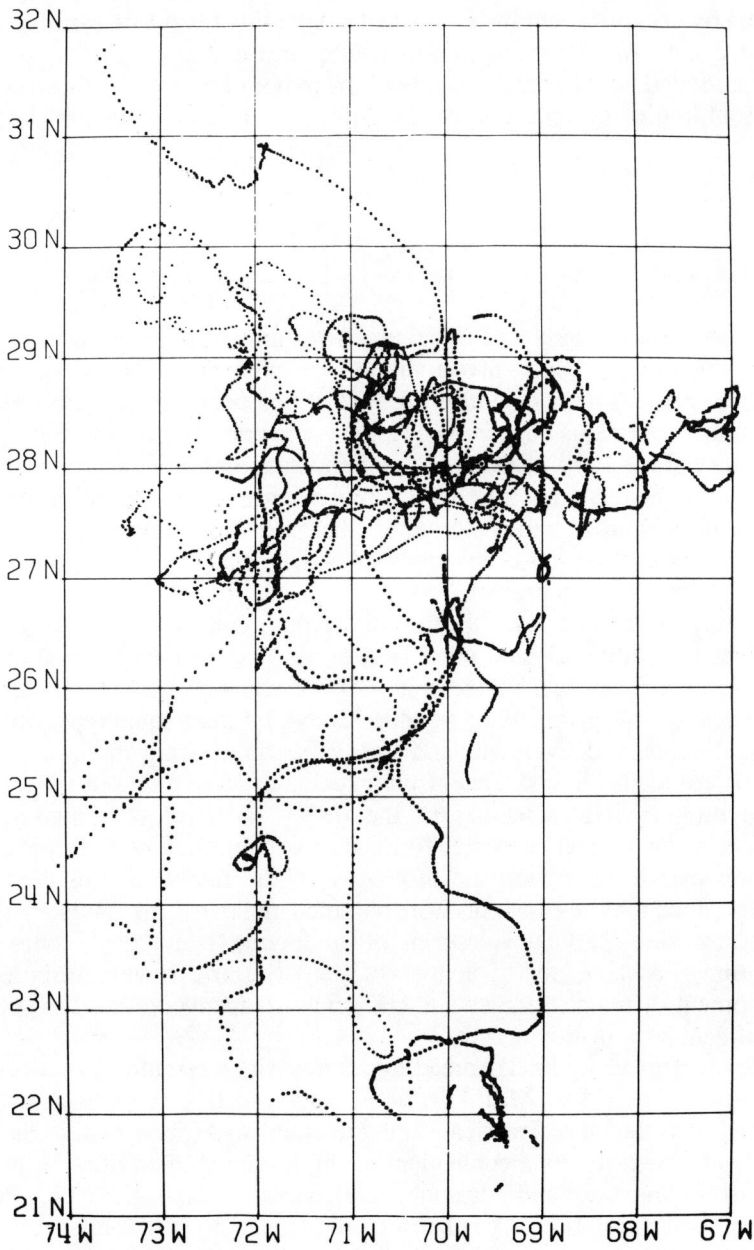

FIGURE 4. MODE experiment. A map showing the 4½ month track of a neutrally buoyant float at depth in the Caribbean.

climate exercised by the oceans. This is constantly evident in countries like ours affected by cyclonic disturbances from the Atlantic; it is as true worldwide. Here again we have to be concerned about long-term climatic changes such as the often-predicted next ice age, and one of the major items in this study must be an understanding of oceanic circulation and its long-term variation.

GEOLOGY OF THE DEEP OCEANS—KEY TO WORLD STRUCTURE

The most striking advance this century in the understanding of world structure has arisen from oceanic investigations by physical methods (particularly magnetic and seismic data) followed by drilling operations in deep water. The concept that continents have moved relatively to one another is not new, but the availability since 1960 of detailed magnetic data on the ocean floor combined with seismic analysis has provided a much more sophisticated and coherent hypothesis which accounts for the changing distribution of oceans and continents, and which has been supported in detail and extended by the results of the Deep Sea Drilling Project.

The outer layer of the Earth, the lithosphere, consists of large rigid plates which move relative to each other (figure 5). The boundary regions between these plates are marked by tectonic activity (earthquakes and faulting) and sometimes by volcanism. Plate boundaries exhibit three main types of motion, that is, divergent, convergent, and transverse or shearing motion.

These are the basic concepts of plate tectonics. Beginning in 1963 with the confirmation by UK scientists of the reality of sea-floor spreading, plate tectonics is now a widely accepted foundation on which many studies of the solid earth are based. Following the discovery of the importance of the oceanic magnetic lineations new insights were obtained into transform faults, earthquake first-motion studies, the development of subduction (convergent) zones and the generation of oceanic crust. Scientists of many countries have contributed to the development of plate tectonics and UK scientists have continued to play a not insignificant role in this process.

Plates diverge along linear spreading centres which commonly occur as major submarine ridges. The Mid-Atlantic ridge is such a ridge and marks the boundary between the American and Eurasian or African plates. Spreading centres, of which this is a convenient example, are studied because it is here that most oceanic crust and lithosphere are generated and begin to be influenced by many ageing processes. The plate tectonics concept, with its implications that the earth is a very active and dynamic heat engine, with continuously changing geography of continents and oceans, has been applied in many fields and is being further developed by continuing research.

A large part of the activity in this field has been funded and carried out by the oceanographic institutions of the USA, but British geophysicists and geologists

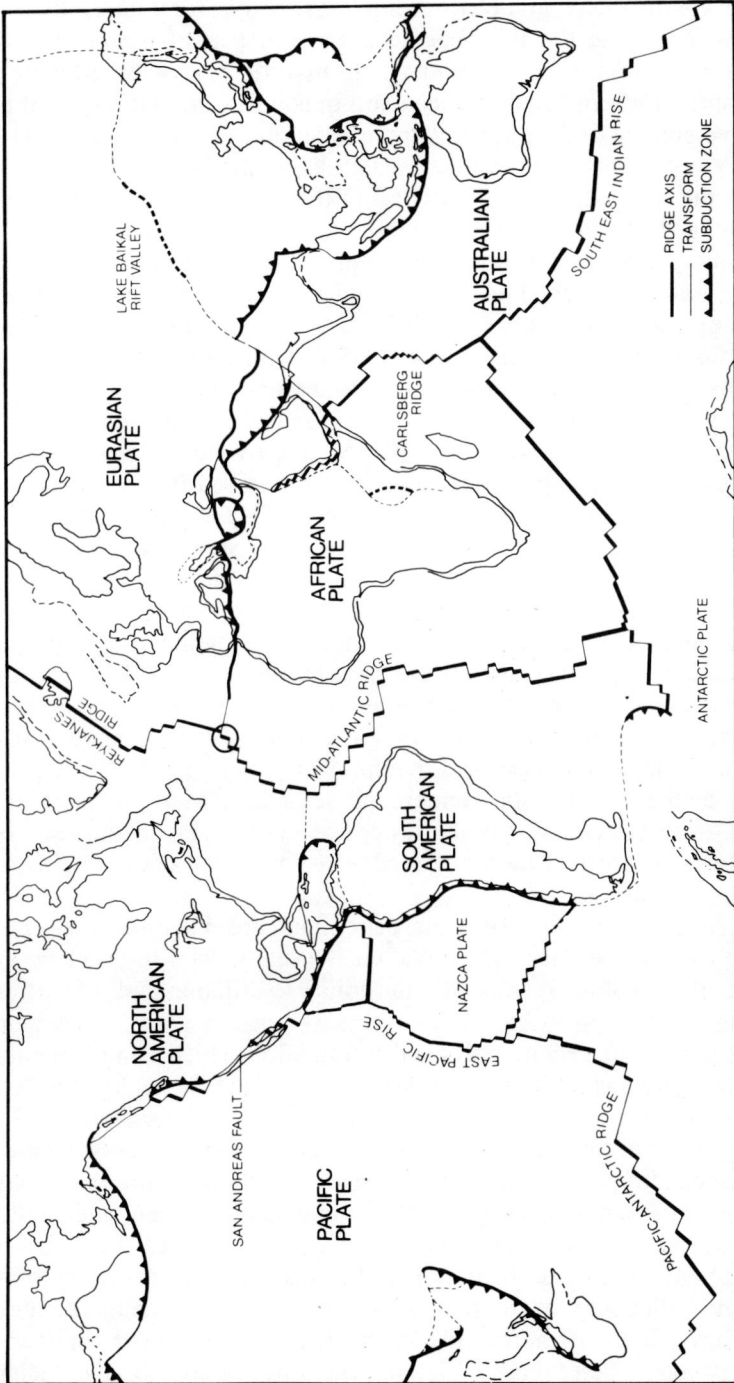

FIGURE 5. The outer layer of the earth, the lithosphere, consists of a few rigid plates which move relative to each other.

have been continuously involved—being responsible for one of the critical concepts which made the magnetic data a major part of the plate tectonics hypothesis, by continuous involvement in the investigation of the structure of the North Atlantic, Eastern Mediterranean and Indian Ocean, and responsible also for the development and geological application of a variety of side-scan sonar systems. A key contribution in this research has come from the ultimate range member of the side-scan sonar family, GLORIA (acronym for Geological Long Range Inclined Asdic), developed for marine geologists at IOS. This low-frequency sonar, larger and more complex than a midget submarine, is able to 'scan' a swathe of seabed up to 27 km wide at an oblique angle and yields a plan view of seabed features. GLORIA emits regular pulses of sound along a narrow beam which strikes the sea-floor. Sound which is backscattered from the sea-floor by suitably rough surfaces is detected and displayed, relative to the time since the pulse was emitted, as white marks against a black background. Scans from successive pulses are displayed side by side so that eventually a sound picture or sonograph, representing a strip of the sea-floor, is built up in the same way as a television picture is made. Repeated passes on a grid system have produced a detailed picture of a large area of the spreading plate boundary southwest of the Azores. The method has also been directed at a study of the Eastern Mediterranean which oceanographers find fascinating because it is here that a young mountain range may be growing. Interpretation of sonar graphs is a fine art, but data from other instruments have been used to check the accuracy of the conclusions and they have been fully confirmed. Despite the spectacular range of data that has made GLORIA the envy of other oceanographers the Mark I version has serious limitations and an improved version (Mark II) has now been developed at the IOS and has undergone sea trials. Mark II has a smaller transducer array system; it is much faster and safer to launch and recover and provides twice as many data, as it is able to scan both sides simultaneously (figures 6, 7, plates II, III).

Geophysical surveys can cover large areas and provide broad concepts, but without other controls their interpretation is usually open to differing views. Development of drilling vessels with the ability to drill hundreds of metres into the ocean floor provided the necessary sedimentary, geochemical, palaeontological and geochronological information which complemented the physical data and has led to general acceptance of the 'new global tectonics', which has made sense of a great deal of formerly disparate geological information. One of the consequences of the plate tectonics hypothesis has been a new understanding of the origin and distribution of the metalliferous deposition of the continents; another has been the development of a new branch of earth science—Oceanic Palæoenvironment—which investigates the chemistry, temperature and biology of the oceans of the past, with their relation to former climates and with a potential for prediction of future world climatic changes. UK scientists have taken a leading role in this research from their application of oxygen isotope determinations to measure past ocean temperatures.

THE MINERAL WEALTH OF THE SEA AND OCEAN FLOOR

An observer standing on the east coast of England, near the Tees or the Humber, the Wash or the Thames, looks out over wide spreads of sand and mud brought down by the rivers, some of it accumulated in the estuaries as shifting sand and mud banks, much of it going on to be deposited on the floor of the grey North Sea. If he moves a few miles inland he will see that the rocks of the coastal belt—the limestones, chalk and other strata—dip down eastwards to disappear beneath these modern sediments. If the observer moved to the Netherlands, and gazed westwards he would again see accumulating modern sediments and might be conscious that older rocks under his feet were again dipping down (westwards) beneath the North Sea. The North Sea has in fact been a gently subsiding sedimentary basin for nearly 300 million years. In it many thousands of metres of sediments have progressively accumulated, these are the erosion products of older rocks of England, Scotland and mainland Europe (figure 8).

Twenty-five years ago the North Sea, like the ocean floors, was largely terra incognita, an area where pure (sometimes wild) speculation about the geology was the rule. Exploration for gas and oil has now made the North Sea perhaps the area best studied in depth in Europe, productive both of scientific understanding and mineral wealth.

The oldest rocks with which we are at present concerned in the North Sea are the Coal Measures, pre-dating the formation of the present sedimentary basin, which were laid down in a wide belt extending from the Irish Sea eastwards to Poland, and which form the floor to the sediments laid down in the sinking basin which we see today. The Coal Measures are important for their large quantity of carbonaceous material, which has been progressively metamorphosed with the increase of temperature and pressure as they have sunk to greater depths, beneath the thousands of feet of later sediments that have accumulated on top of them. This loading has made the coals progressively harder and more anthracitic, and, more importantly, the process has generated very large quantities of methane which have migrated upwards through complex pathways into the overlying newer sediments—our valuable 'North Sea Gas'.

Fortunately for us, the sedimentary fill of the North Sea basin began with a desert period in which broad spreads of dune sands were blown into the area, and after this phase the North Sea basin (then including northeastern England and northern Germany) was occupied by a series of great evaporating basins, in which marine water was progressively concentrated so that it deposited first carbonates of magnesium and calcium, then sulphates (anhydrite and more complex salts of calcium, magnesium and potash) and chlorides. Sodium chloride (halite) accumulated to thicknesses of thousands of feet, and the ultimate concentration produced beds of potash salts, long exploited at Stassfurt in Germany, now being developed in East Yorkshire as a valuable source of fertiliser.

This complex combination of circumstances, with carbonaceous Coal Measures progressively losing methane as they were buried deeper and deeper,

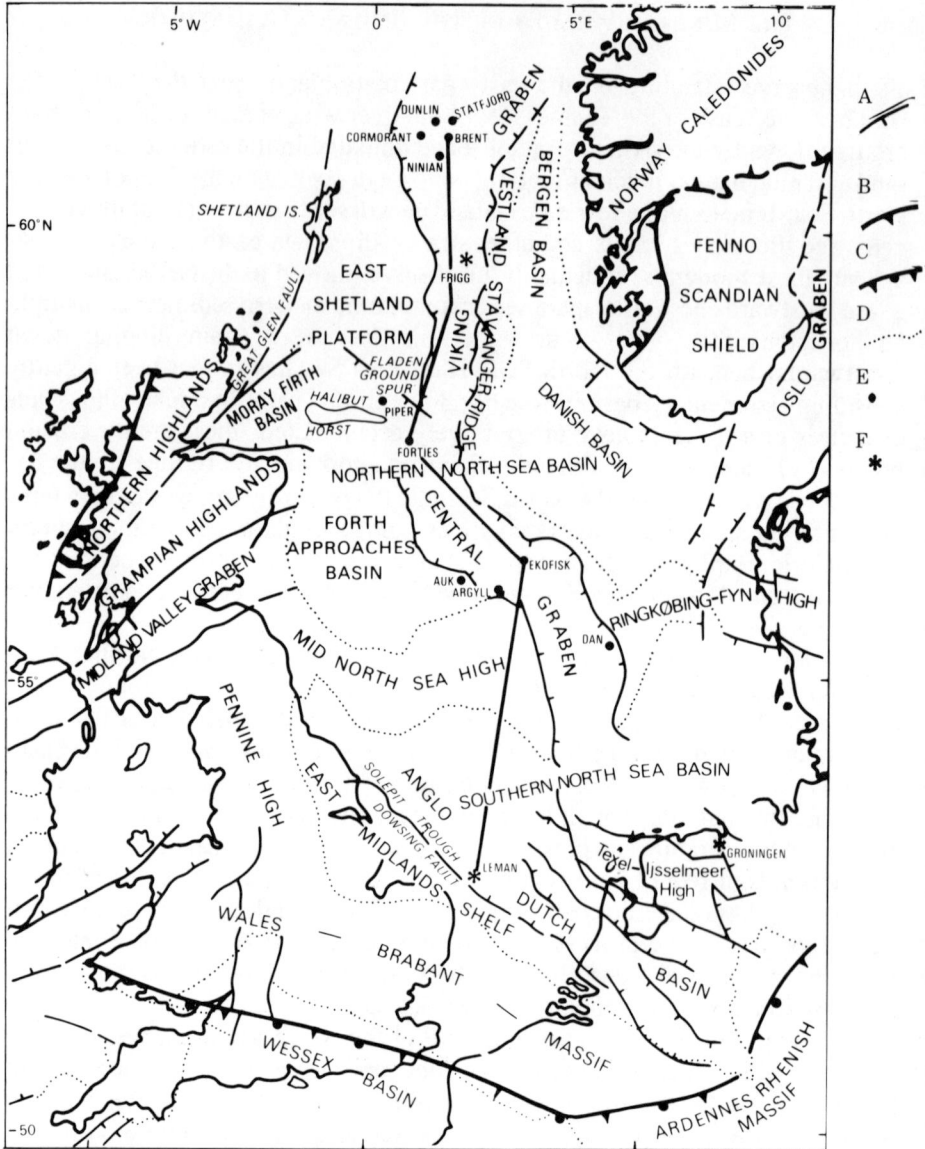

FIGURE 8.   Major structural elements of the North Sea. A, fault; B, Caledonide front; C, Variscan front; D, form line; E, oilfield; F, gas field.

with porous sands to provide reservoirs for gas, and—no less important—with an overlying evaporite sequence to provide a very effective seal on the reservoir sands and prevent the loss of the methane to atmosphere, is responsible for the availability of the major energy source of the southern North Sea (figure 9).

This was only the beginning of the long story of sedimentation on the North Sea, and of the development of its hydrocarbon wealth. The 'evaporating dish' phase (the Permian) was followed by a long desert period in which deposition of red sands and muds dominated (the Trias), after which the sea flooded into the basin, inaugurating the marine regime which still persists after 150 million

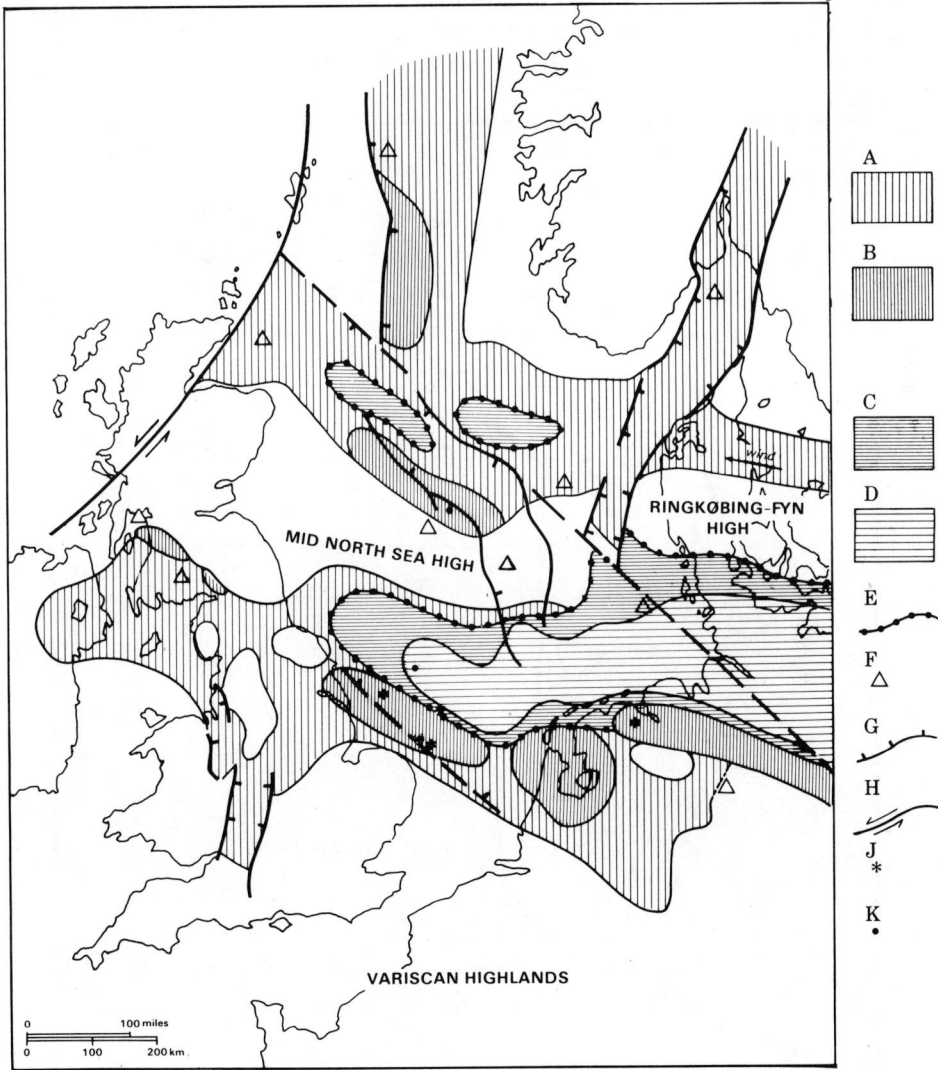

FIGURE 9. Lower Permian palæography. A, Leman sandstone; flood plains and wadi deposits carbonate cemented; B, Leman sandstone: dune sands; C, Silverpit formation: shoreline deposits: dolomite, anhydrite, silts, clays (sabkha); D, Silverpit formation: salts in periods of drought; E, fluctuating shoreline; F, volcanic centres; G, fault; H, wrench fault; J, gas field; K, oilfield. (*After* K.W. Glennie.)

years. Into these marine waters sands were washed from the Scottish Highlands and Scandinavia; at other times dark organic muds accumulated, as indeed they do today. The greatest thicknesses of these sediments were found in the northern North Sea, where more than 17 000 feet of them has been drilled in some places. Again deep burial resulted in higher temperatures and pressures and the consequent chemical processes; the organic muds generated liquid hydrocarbons which migrated into the nearest reservoirs—in this case the marine and estuarine sands of Jurassic and Tertiary age—to provide the major oilfields now being developed off Scotland and the Shetlands (figure 10).

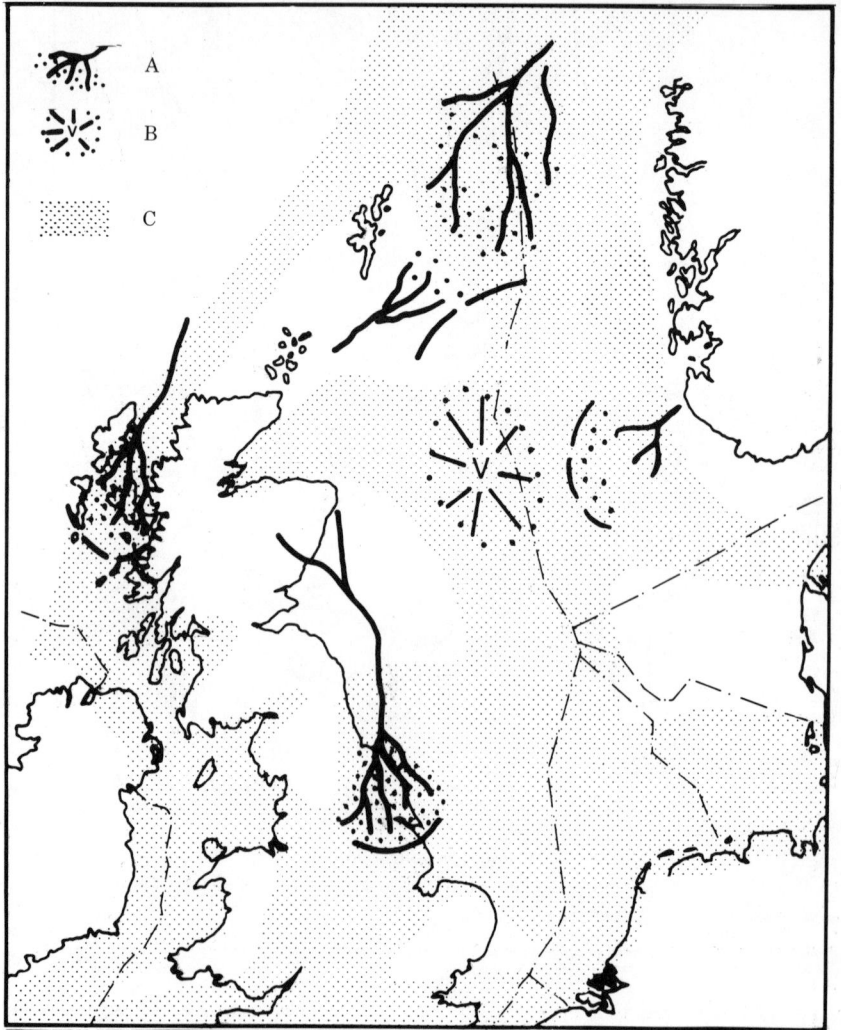

FIGURE 10.   Middle Jurassic deltas. A, deltaic sand development; B, volcanic complex; C, extent of Middle Jurassic deposition.

What is the next phase of exploitation of the sea's mineral resources? Already exploration is extending beyond the continental shelves, in Northwest Europe and elsewhere in the world. Research organized internationally by oceanographic institutes has made a major contribution to geological knowledge of continental slopes, rises and the ocean floor. Parts of the continental slopes, at least, have very thick sediments and a potential for providing further hydrocarbon resources. The deep ocean floor is a different case—sediments are relatively very thin, but vast areas are scattered with manganese nodules (figure 11, plate IV), segregations rich not only in manganese but also nickel, silver and other metals. Investigations of the mid-ocean ridges have lately found hot spring activity and metalliferous desposits; hot metalliferous brines have also been found in deep pockets in the Red Sea.

Development of these deeper resources is a matter for the future, but the research which we do now is a necessary preliminary to making them available for the long-term benefit of mankind.

ACKNOWLEDGMENTS

This outline account of some of the highlights of the marine activities of component institutes of the Natural Environment Research Council incorporates information prepared for the Jubilee Exhibition organized by the Institute of Oceanographic Sciences and the Institute of Geological Sciences, and is illustrated by diagrams exhibited on that occasion. The author is indebted to the contributors and other colleagues for helpful comments on the draft version of this account.

CONTRIBUTIONS OF SCIENTIFIC DISCOVERIES TO INCREASES IN AGRICULTURAL PRODUCTIVITY

G.W. Cooke, F.R.S. and Sir William Henderson, F.R.S.

Agricultural Research Council, London

Introduction

British agriculture is a success story and the Queen's reign has witnessed the most rapid transformation that our agriculture has ever experienced. Before the 1939–45 War we depended for roughly two-thirds of our food on imports from overseas. Now two-thirds of the kinds of food that can be grown in our climate, or just over half of our total food requirements, are home-produced on less land and for a larger population.

Crops

The yields of all crops have been greatly increased by the application of the results of scientific research. Plant breeders have produced varieties of greater yield potential, more resistant to disease, which are now well nourished by fertilizers. The chemical industry has provided materials to control the weeds, pests and plant diseases which formerly took a far greater toll. In some crops increased yield and increased area have together resulted in great increases in production: barley is an example where output has increased ten-fold. With other crops we already grew as much as we needed but the area required has been dramatically reduced because yields have risen; the present output of potatoes requires only 60% of the area which it would have needed 25 years ago. Yields of crops grown under glass, such as tomatoes and cucumbers, have been increased so that more is grown for the same expenditure on fuel.

Livestock

The numbers of livestock kept are now greater than 25 years ago. Breeders have provided genetically improved animals, and as these are more efficiently fed and protected from disease they have responded by increased production per animal. Imports of animal feeds have not increased, the extra food being found from improvements in output from our arable and grassland. We produce

53

practically all the eggs, poultry meat, pork and all the fresh milk consumed in this country. Well over 80% of the beef and veal we eat, and half the mutton and lamb, is home produced.

Land

Although production has increased, the land available for agriculture has diminished as areas are diverted from farming to urban and industrial development, roads and forestry. It is estimated that during the last 25 years 0.2% of the area of agricultural land has been taken each year for these non-farming purposes.

Labour Productivity

The full-time farm workers in England and Wales have declined from 534 000 in 1952 to 171 000 in 1977; (in the latter year there were also 220 000 farmers). This change has been possible through the great increases in mechanization which have given British agriculture the reputation of being the most highly mechanized in the world. Adequate power and improved farming techniques have contributed, through better and more timely cultivations, to increased crop yields. They have also contributed to labour productivity which increased at 4.5% per year in the ten years 1964/65 to 1974/75. Putting the gross volume of production per worker at 100 for 1968/69 to 1971/72, the corresponding figures were 76 in 1965/66 and 133 in 1974/75.

PROGRESS IN FEEDING OURSELVES

The great improvements in agricultural productivity have lessened our dependence on imported food and in 1974/75 we produced 54% of all that we ate; of the kinds of food that can be produced here, UK farmers provided 68%. Some details of changes in the last 40 years are given in table 1.

TABLE 1

Home Production of Foods as percentages of totals needed

	Pre-War (1936–39)	1953/54	1964/66	1974
Wheat	23	41	46	68
Barley	46	67	97	92
Potatoes	96	98	96	95
Beef and veal	49	66	69	78
Mutton and lamb	36	35	43	54
Pork	78	88	98	99
Bacon and ham	29	43	36	46
Poultry meat	70	–	98	99
Eggs	61	80	98	98
Milk (liquid consumption)	100	100	100	100
Cheese	24	38	43	64

In interpreting these figures several facts are important: the population has increased by 10% since 1952; incomes have risen so that most people are able to purchase an adequate diet; in 1952 the aftermath of war-time rationing on the availability of food still affected supplies and purchasing patterns. Scientific developments have also helped to provide products of higher quality (that is, more suited to particular purposes). For example, twenty-five years ago flour used for bread-making contained only about 10% of English or European wheat; much of the British wheat produced was used for animal feed and for making biscuits. As a result of new processes for baking bread and of the breeding of wheat varieties that suit these, about half of the grist is now from English wheat and the proportion may, in future, rise to 70%.

In addition to producing an increased proportion of our own food we send much food and feed to other countries; these exports were worth £221M in 1960, £484M in 1970 and £1367M in 1975. In real terms this amounts to an increase of two and one-third times in food exports in the 15 years.

Research and Development

The rapid changes over the last 25 years in farming patterns, and in productivity, have been the result of the application of discoveries in science and engineering. Most of the agricultural research done in Britain is in the Institutes and Units for which the Agricultural Research Council (ARC) is responsible for advising on programmes. The Council has eight Institutes and six Units under its direct control; 16 other Institutes in England and Wales are grant-aided by the Council. In Scotland eight Institutes are financed by the Department of Agriculture on the advice of ARC. Other research relevant to agriculture is done in the Universities and some is directly financed by Government Departments. Advice on research programmes is provided by the Joint Consultative Organization which is jointly sponsored by the Ministry of Agriculture, Fisheries and Food (MAFF), the Department of Agriculture and Fisheries for Scotland (DAFS) and ARC. Advice to farmers on the application of scientific work done to improve agriculture is the responsibility of MAFF and DAFS. In England and Wales MAFF established the National Agricultural Advisory Service (NAAS) in 1946; in 1970/71 NAAS was re-organized as the Agricultural Development and Advisory Service (ADAS). In Scotland the responsibility for development and advice is with the three Scottish Colleges of Agriculture. The close associations which have been established between research workers and those responsible for development and advice must have much credit for the speedy adoption by farmers of the results of scientific discoveries. Examples of successful research, and its application to practices, are given in this paper.

IMPROVEMENTS IN CROP PRODUCTION

Yields of most arable crops have been substantially increased since 1952. Table 2 shows examples for a few important crops.

TABLE 2

Change in Average Yields of Crops, 1939–1974 in UK

	1939	1952	1963	1974
	Yields (tonnes per hectare)			
Wheat	2.3	2.9	3.9	5.0
Barley	2.2	2.6	3.5	4.1
Oats	2.1	2.4	2.8	3.8
Potatoes	19.1	19.8	21.6	31.6
Sugar beet	25.9	26.1	31.4	(36.1)*
Tomatoes (under glass)	84.0	84.0	86.0	123.0

*average (1969–73)

Most of these improvements in crop yields are the result of advances in several branches of agricultural science. Plant breeders have the basic responsibility for seeking new varieties with improved characteristics such as higher potential yield and increased disease resistance. The greater potential of the new plant material can only be achieved by adequate nutrition and by keeping the crops free from pests, diseases and weeds. Research in agronomy seeks to define the ways in which physical, chemical and biological factors interact to affect plant growth, and to suggest agricultural practices likely to give optimal yields of newer crop varieties. As a result of such work, most arable crops in the UK receive an adequate supply of nutrients, whilst crop protection chemicals are used extensively by farmers to give better control of pests and diseases.

Improvement of Wheat by Plant Breeding

The increased yields of cereals shown in Table 2 have been mainly due to improved husbandry techniques involving increased use of nitrogen fertilizers and to the breeding of new varieties. These are higher yielding, more resistant to disease and have shorter and stiffer straw, so that they can make full use of the extra fertilizers without lodging. The winter wheat Maris Huntsman, introduced from the Plant Breeding Institute in 1972, was 10% higher yielding than all other varieties then available in Britain and is now grown on about 40% of the wheat land. It has been followed by shorter strawed varieties with even greater yield bred at the same Institute; Maris Hobbit and Maris Statesman are now available to farmers and other improved varieties are in trials.

Breeding for Quality

Much of the wheat grown in Britain is used to feed animals, but the proportion used in bread-making flour has increased from about a fifth in the 1960s to about half in 1976. Baking quality is largely dependent on the quantity and quality of the flour protein though it may be spoiled if the weather at harvest time is bad. The increased use of British wheat has been made possible by three factors: the introduction of a new baking procedure, the Chorleywood Bread Process devised at the Flour Milling and Baking Research Association, in which satisfactory loaves can be made from flour of rather lower protein content; the use of more nitrogenous fertilizers, especially during the later stages of crop growth; and the breeding of varieties, such as Maris Freeman, with better milling and baking quality.

Breeding for Resistance to Disease

The yield of cereal varieties may be seriously reduced by diseases. The best way of preventing such losses is to breed resistant varieties. For example, resistance to the eyespot fungus (*Pseudocercosporella herpotrichoides*), which used to cause severe losses to wheat crops grown successively in the same field, has been introduced by breeding, so that most varieties now cultivated in Britain are resistant.

Breeding for resistance may be difficult if the diseases become adapted to new varieties through the development or natural selection of strains capable of attacking them. This problem is most serious with diseases such as yellow rust (*Puccinia striiformis*) and powdery mildew (*Erysiphe graminis*) which are spread by airborne spores. Breeders are therefore attempting to breed varieties with durable resistance, which it is hoped will not break down to new strains of the pathogen. Some varieties, such as Maris Huntsman, show reasonable levels of resistance to yellow rust and powdery mildew although pathogen strains capable of overcoming some of their resistance have arisen.

Crop Nutrition

Fertilizers are essential in modern intensive agriculture because they build the large yields which we must have to repay large capital and recurrent costs. By introducing extra plant nutrients into cycles of growth and decay, fertilizers eliminate one natural limitation to crop production—the supply of plant nutrients from soil. ('Natural' soils rarely supply enough of all the nutrients that crops need for maximum yield.)

Although the fertilizer industry was established in Britain 130 years ago, it is only in the post-war period that it has developed new methods which give a large capacity for production at costs which have risen less rapidly than prices of most other products which the farmer uses. Table 3 shows how the amounts of plant nutrients supplied by fertilizers have changed during the last century. In 1975/76 the industry supplied fertilizers and liming materials for British agriculture which cost £344M.

TABLE 3

Amounts of Nutrients (as elements) applied as Fertilizers
in the UK 1874–1975

	N	P	K
	thousands of tonnes		
1874	35	30	3
1939	61	76	63
1952	184	123	145
1965	574	212	358
1977	1093	177	340

Over the last 30 years cereal crops have received a steadily increasing amount of fertilizers to support the increased vigour and higher yields of successive new varieties. In this period, during which national average wheat yields have doubled, the average application of nutrients supplied by fertilizers to cereals in Eastern England has trebled. Potato yields have also greatly increased over the last 30 years from 19 t/ha to 32 t/ha, due primarily to the increased use of nutrients and to improved cultural practices. Recent investigations show that, when even larger fertilizer dressings are applied, normal yields of 50 t/ha should be possible and 80 t/ha have been recorded.

At present we believe that sufficient fertilizers are being applied to our arable crops. The major part of our grassland will yield better when it receives more nitrogen fertilizer. However, progress in fertilizing grassland is inevitably slower than in fertilizing arable crops. Grass produced by fertilizers must be matched to the needs of the extra livestock needed to consume it, and the way they are managed; these adjustments to animal farming systems take longer than adjusting an arable system to harvest larger crops.

Control of Diseases

Leaf diseases

Foliar diseases are important causes of yield loss in cereals, and of these barley powdery mildew (*Erysiphe graminis*) is the most damaging in UK. In 1968 this disease caused losses in England and Wales estimated as 1.1 million tonnes. Resistant varieties of barley have been only partially successful in preventing losses. In the last ten years, however, a number of very active new fungicides have been developed: they have systemic properties, acting by transport through the plant, and are now used to control powdery mildew. In

1976, 42% of the crop was treated with fungicide at a cost of £5–6M which prevented a loss estimated at 350 000 tonnes of barley or £29M. Recently, strains of powdery mildew tolerant to some fungicides have emerged. Management systems for the pathogen based on the use of both fungicides and resistant varieties of barley are likely to provide the best solution at present.

Virus diseases

Virus diseases can cause heavy losses in crops, particularly cereals and potatoes. They can be controlled prophylactically, by reducing the population of vectors such as aphids, or by two therapeutic techniques which can be applied to vegetatively propagated plants. First, by excising the meristem which, when grown in artificial culture, produces a plantlet free from virus, and second, by heat therapy during which plants are kept at 35°C for two to three weeks. For example the future of the King Edward potato was threatened because of its susceptibility to virus attack, but research at Rothamsted Experimental Station with meristem cultures enabled this variety to be grown free from the damaging paracrinkle virus. Both meristem culture and thermotherapy have found worldwide application in the eradication of viruses from crops, with incalculable benefits.

Control of a Pest—Potato Cyst-nematode

In the past, the potato cyst-nematodes (*Globodera rostochiensis* and *G. pallida*) reduced potato yields in the UK by 10% or 500 000 tonnes per year. Until recently nematode populations could only be controlled by resting infested land from potatoes for periods up to ten years, or even more.

Nematicides

Recently discovered nematicides of the oxime carbamate type (aldicarb and oxamyl) now enable any variety of potato to be grown in heavily infested land. Experiments by Rothamsted have shown that they can prevent loss of yield and reduce nematode increase, on average, by 90%.

Breeding of resistant varieties

Breeding for resistance to potato cyst-nematode started in 1952 using resistant material which included five clones of cultivated tetraploids from Bolivia and Peru. These Andigena potatoes were crossed with European Tuberosum cultivars to obtain resistant varieties with a high yield. The most important of these is the maincrop Maris Piper, introduced by the Plant Breeding Institute in 1966 and occupying 26 000 hectares in 1976; Pentland Javelin, introduced in 1969, is a resistant early variety.

Resistant varieties cause larvae to hatch out from cysts which then invade the roots but contrary to their behaviour in the roots of susceptible varieties, the larvae are unable to develop into new cysts (mature females). Hence growing resistant varieties reduces the cyst-nematode population in the soil instead of increasing it.

By using these control measures (nematicides and resistant varieties) annual yield losses of potatoes amounting to £50M can be prevented and the national potato crop can be grown on less land.

Chemical Control of Weeds

The first herbicides to be developed on a large scale were those for the control of dicotyledonous weeds in cereals. These were taken up extensively by British farmers and rapidly overturned the concept of crop rotation for weed control; cereals changed from being a fouling crop to a cleaning crop. Even those weeds which were wholly or partly resistant to the widely used phenoxyacetic acid derivatives proved susceptible to the subsequently developed phenoxypropionic acid derivatives and a whole range of other active compounds which have continued to provide control of dicotyledonous weeds in cereals at moderate cost. Chemical industry has similarly provided herbicides for use in all major crops, including the rowcrops traditionally cleaned by inter-row cultivation and hand hoeing. The Weed Research Organization (an ARC Institute) has played a major part in these developments.

The trends away from crop rotation towards more intensive cereal cropping and from mouldboard ploughing towards a reduction in cultivation have favoured the increase of annual grass weed species. Wild-oats (*Avena fatua* and occasionally *A. ludoviciana*) are now the most important weeds of cereals throughout England and parts of Scotland. Black-grass (*Alopecurus myosuroides*) can be just as serious where it occurs. The most recent research has shown that chemical control must be closely integrated with good agronomic practices to achieve maximum efficiency. In modern practices of weed control herbicides have become the essential cornerstone, rather than soil cultivation.

Inputs to Agriculture from Chemical Industry

Chemical industry has played a vital role in the agricultural revolution of the last 25 years by supplying the materials for the nutrition of crops and their protection from weeds, pests and diseases and for the improvement of animal health. The agrochemicals industry has expanded rapidly; output in two recent years was valued at: UK sales, £64M in 1974, £97M in 1976; exports, £57M in 1974, £94M in 1976. Of the total in 1976 £57M was for herbicides, £13M for insecticides and £9M for fungicides, all used in agriculture and horticulture. The industry also supplies the chemicals used for weed, pest and disease control in industry and in private households and gardens. Seed-treatment products used in agriculture were valued at £4M in 1976 and UK sales of chemicals for control of animal ectoparasites were worth £2M.

IMPROVEMENTS IN LIVESTOCK PRODUCTION

The total receipts from UK agriculture were valued at £5456M in 1975/76. Of this, farm crops provided £1211M and horticulture £551M; livestock and livestock products contributed £3511M – nearly two-thirds of the total.

We have about 50% more cattle, 30% more sheep, 75% more pigs and nearly 50% more poultry on our farms than in 1952. Increases compared with pre-war figures are even more dramatic as Table 4 shows.

TABLE 4

Changes in Livestock Populations in UK 1939–1974

	1939	1952	1963	1974
	(millions in June of each year)			
Total cattle and calves	8.9	10.2	11.7	15.2
Sheep and lambs	26.9	21.7	29.3	28.5
Pigs	4.4	5.0	6.9	8.5
Poultry	74.4	95.0	112.2	139.7

Although numbers of stock have increased, the extra mouths have been fed from the produce of our own farms. Because of improved yields of cereals and the provision of high quality feed from scientific management of our grassland, we need to import less feeds than in pre-war days. Intensive production from pigs, poultry and dairy cows requires high-protein diets and traditionally imports have supplied much of their needs. Current research on protein-rich crops aims to lessen our dependence on these imports even further.

Livestock have been improved by breeding programmes which developed animals that make more efficient use of feed. Research has also led to methods of feeding and management that improve the ratio of conversion of feed to meat, milk or eggs. The net result of these advances, and of improved animal health, has been a considerable increase in yields without commensurate increase in costs. Changes in annual milk yields (in England and Wales) and egg production (in UK) show these trends:

	Pre-war	1954/55	1966/67	1973/74
Milk, litres/cow	2546	3069	3614	4000
Eggs, per hen	144	167	207	225

Examples of the scientific work which has led to these improvements in livestock production are given in the following sections.

Animal Breeding

Improvements in reproductive performance have come from two approaches, genetic and physiological. The geneticist (or breeder) selects and breeds from animals of higher productive potential, whereas the physiologist seeks to understand, and ultimately to control, the reproductive mechanisms of individual animals.

Dairy cows are our most important farm animals and the past 25 years have seen considerable changes both in the genetic constitution of the national herd and in its management. Physiologists and geneticists have contributed to this change by devising, and applying, techniques of artificial insemination (A.I.). Over 70% of our dairy cattle are now bred by A.I. and the technique has facilitated the selection of sires. A.I. has been in general use since World War II but the technique of storing bull semen for long periods without deterioration in liquid nitrogen ($-196°$) has only been available since 1952. Individual 'doses' of semen are now usually frozen and stored in plastic tubes ready for use. Since 1970 more than half a million doses have been sent overseas. An individual bull may supply over 40 000 doses per year.

Dairy cows produce all the fresh milk consumed in this country but also contribute much of our beef. Because of its 'dual purpose' the Friesian breed has become dominant in these islands. Selection for output has raised the average milk yield per cow per year by 50% since 1952. A recent feature of breeding in beef herds has been the introduction of 'exotic' breeds, for example the French Charollais. Performance testing and progeny testing, in pigs and sheep as well as in cattle, has replaced the assessment of individual animals on appearance and their selection by intuition. The methodology of selection has become an important branch of applied genetics.

The widespread use of A.I. in cattle to exploit the genetic potential of the male was largely instigated by the late Sir John Hammond and it was he, some 25 years ago, who introduced the concept of 'artificial pregnancy' (embryo transfer or ova transplantation) to exploit the genetic potential of the female. The ovulation and fertilization of twenty ova or more can readily be induced in cattle and it has been shown that these can be transferred to other cows whose reproductive cycles coincide with that of the donor. This synchrony can be achieved by preliminary treatment using prostaglandins and effort is now concentrated on recovering the embryos without recourse to anaesthesia or surgery. Such techniques have great potential for speeding the multiplication of superior animals.

A single injection of the same prostaglandin analogue given to sows 2–3 days before the expected date of farrowing results in parturition about 26 hours later. Therefore farrowings can be confined within normal working hours, conducive to more careful supervision. Comparable manipulation may become practicable in sheep husbandry through experiments with a corticosteroid analogue to induce parturition.

When it was discovered that the concentration of progesterone in the milk of a pregnant dairy cow reflected that in the blood, a test for pregnancy relatively early in gestation became possible. The analysis was automated and the system field-tested, and a routine service was quickly established at the Milk Marketing Board's laboratory.

Productivity of breeding herds of pigs has been raised by 'early weaning' methods which, combined with selection for prolificacy and 'mothering' qualities, has raised the number of pigs weaned per sow per year to about 22 in the more efficient herds. At the same time, the advantages of cross-breeding in litter size and growth rate have been exploited; relatively few pure-bred herds are now maintained to retain the breed characters. This specialization has been aided by the concentration of the bulk of the pig population into ever fewer herds, with the increased sophistication of management that this trend encourages.

Advances in sheep husbandry have been less dramatic than those in pigs and cattle, mainly because of the diffuse nature of the industry, but selection has significantly raised the prolificacy of lowland sheep. Hill sheep present special problems; an increased lambing percentage, for example, is all too easily nullified by increased mortality.

Animal Nutrition

Progress with nutrition of farm animals has been continuous rather than spectacular. The assessment of feed value was fundamentally changed by the concept of 'metabolizable energy' in the early post-war years, replacing 'starch equivalent' in calculating feed allowances. This concept was derived from experiments in direct calorimetry, which measured heat produced and oxygen, carbon dioxide and methane exchanges involved, on different types or quantities of feed.

Cattle and sheep are of special importance for their capacity, as ruminant animals, to feed on grass. The biology of the rumen has therefore been intensively studied, particularly the role of the rumen bacteria in breaking down cellulose and the subsequent availability of the metabolic products, including the bacterial population itself. The demands of early lactation in dairy cows cannot be met from what the cow can extract from her feed during that time. She draws on reserves in her own body, and her milk yield is higher, and better maintained, if her reserves are built up by feeding concentrates in late pregnancy, a process which its principal innovator called 'steaming up'. Research with rumen-fistulated animals has done much to explain the relative contribution of various dietary constituents to this effect, and has related their consumption to the composition as well as to the quantity of milk. Such findings are taken into account in formulating prepared concentrates. The complexity of the mixtures, the variety of substances required in small amounts, and the demands of different purposes in any one species, have led to a substantial industry supplying the farmer with complete or supplemental diets to meet a wide range of circumstances.

Through advances in research techniques, the fate of dietary constituents can be studied during passage through the gut. Such experiments do not lead to dramatic changes in husbandry but to a continual adjustment as the results are equated with practical situations—involving, for example, the availability or cost of various types of forage or feed. In contrast, the use of antibiotic additives to cattle feed may be said to have passed through a phase of enthusiasm to an era of discretion, and the same is true of the use of anabolic steroids to promote growth.

The outstanding example of the trend towards intensive stock farming is in poultry, especially within the meat section which comprises broiler and turkey production. The table chicken has been transformed from a luxury to the cheapest animal protein available. Poultry, like pigs, are monogastric and require a 'high grade' diet; the cost of feeding stuffs more than doubled between 1950 and 1974. The incidental costs of transport and selling have risen correspondingly, but the retail price of the product has not risen. This remarkable increase in efficiency has largely been due to an improved feed conversion rate—from 3.2 to 2.2:1—together with an increased growth rate, birds now taking 7 to 8 weeks to attain the weight formerly reached in 10 to 12 weeks. Both selective breeding and nutritional research have contributed to these improvements. The trends toward larger production units and increasingly sophisticated marketing methods have been mutually accelerative and consumption rose from one million chickens in 1953 to 300 million in 1975. An over-riding factor in the 'mass production' of poultry (as with intensive production of other animals) has been the control of infectious diseases, such as coccidiosis and fowl paralysis (Marek's disease), partly by hygiene but also by the development of vaccines to control outbreaks.

Animal Health

Improvements in animal health have involved the eradication of a number of major diseases of farm livestock and the more effective control of others. In some examples, discussed in the following paragraphs, research done during the last 25 years has made a significant contribution. Much of the research has been done by the Agricultural Research Service but the responsibility for field programmes of control and eradication has been with the State Veterinary Service of the Ministry of Agriculture, Fisheries and Food.

Food-and-mouth disease outbreaks, sometimes numerous, occurred in every year up to 1960. Afterwards the incidence diminished, with no outbreaks in 1963 and 1964, until the dramatic episode of 2364 outbreaks in 1967–68. Since June 1968, the disease has not occurred in the UK. Research has played a vital part in the development of diagnostic tests, in establishing the epidemiology of the disease, and in developing methods of control. Improved methods are used to produce vaccines which are *used in other countries* to keep down the level of disease in livestock populations from which infection might gain entry to this country. The World Reference Laboratory for foot-and-mouth disease is established at the Animal Virus Research Institute, Pirbright.

Brucellosis is being progressively eradicated from British cattle, for example, a survey in England and Wales at the end of June 1976 showed that almost three quarters of the dairy herds were accredited (continued freedom from reactors to the diagnostic test) and that in South Wales the percentage was more than 90. Two types of vaccine have been used for control of the disease: the strain 19 vaccine was very thoroughly examined in the early stages of its use at the Institute for Research on Animal Diseases; the 45/20 vaccine had its origin at Wye College and was further developed at the Animal Diseases Research Association's Moredun Institute. All the strain 19 vaccine used in this country is produced at the Central Veterinary Laboratory.

Mastitis has caused losses of more than £30M a year; it is now much better controlled by a system of hygiene developed by the National Institute for Research in Dairying in collaboration with the Central Veterinary Laboratory.

Swine Vesicular disease appeared in pigs in 1972 and caused particular concern because of features resembling foot-and-mouth disease. Work had already been done on this virus by the Animal Virus Research Institute and the separate nature of the infection was rapidly established, the virus was isolated and its epidemiology unravelled. This made it possible to develop methods of controlling and eradicating the disease.

Marek's disease in poultry (formerly called fowl paralysis or neurolymphomatosis) has caused losses for more than 50 years. With the growth of modern intensive systems of poultry management the disease increased in severity and by 1970 had become the major cause of mortality in poultry, responsible for annual losses estimated at £10M. Since 1970 the disease has been controlled by vaccination, which has reduced the incidence of Marek's disease so that, today, it is no longer a major problem to the poultry industry.

Marek's disease is a tumourous condition characterized by lymphoid tumours in several organs, particularly in the ovary, and which has an unusual predilection for peripheral nerves. The first significant advance was the successful experimental transmission of Marek's disease, achieved in 1962 at Houghton Poultry Research Station, by the inoculation of young chicks of a susceptible strain with blood cells from affected birds. The disease was shown to be infectious and contagious, but its cause remained unknown until 1967, when the application of modern tissue culture and electron microscopy techniques led to the discovery of a herpesvirus as the causal agent. Work at Houghton and in the USA led to the development of commercial vaccines which are now used routinely to vaccinate all day-old chicks used for egg-laying and breeding.

In addition to the agricultural importance of Marek's disease, it is attracting increasing biomedical interest, for it provides a model system for the study of tumour-producing herpesviruses which has relevance to certain neoplastic diseases in man. The Marek's disease vaccine represents the first successful development of a commercial vaccine against a neoplastic disease, and studies on this and derivative experimental vaccines point a possible way to the development of similar vaccines for use in human medicine.

Tuberculosis control became feasible as a result of work done in the 1930s. The area Eradication Plan was introduced in 1950; in that year there were 55 000 attested (i.e. tuberculosis-free) herds and 5000 affected cattle were slaughtered.

Ten years later the campaign had covered all of Great Britain; there were 243 933 attested herds and only 28 cattle had to be slaughtered. The campaign was made possible by the research at Central Veterinary Laboratory which developed the purified tuberculins needed to test cattle.

Respiratory and enteric infections. With the elimination of the major diseases, these infections in young livestock are among the principal causes of animal production losses. Research at the Institute for Research on Animal Diseases and at the Animal Diseases Research Association has led to a greater understanding of their multifactorial aetiology with the identification of the significance of various viruses, mycoplasmas and bacteria as pathogens. Work of importance to comparative medicine has been the discovery of rotaviruses as the cause of similar infantile diarrhoeas in calves, piglets and babies.

Production disease. This term is used to describe the nutritional and metabolic disorders which often occur in dairy cows as a result of imbalances in nutritional input relative to the needs of production. A period of especial risk is when the metabolism of the cow is switching from supplying the needs of the unborn calf to providing for the demands of lactation. Work at the Institute for Research on Animal Diseases has provided analytical procedures, data banks for their use in monitoring herd performance, and schedules for the improved prevention of such disorders as parturient hypocalcaemia, acetonaemia, hypomagnesaemia and other deficiencies, some associated with infertility and premature decline in milk yield.

THE FUTURE

In 1975 a White Paper 'Food from Our Own Resources' (Command 6020) was published in which the Government looked "to the agricultural and food industries, with their fine record of past achievement, to work with them in bringing about an expansion of economic agricultural production in the interests of the nation". The White Paper stressed the importance of research and development in achieving these aims. Reports from the Joint Consultative Organization have indicated the topics which must have priority in research if British agriculture is to continue to increase productivity by technological improvement. A few of these recommendations for future work are outlined below.

Improved plants and animals are needed to raise both output and quality of produce. The impetus in plant breeding must be maintained to produce varieties with higher potential for yield of better quality and with improved resistance to diseases and pests. Scientific work on animal genetics must be exploited to develop superior animals which convert their food more efficiently to muscle protein. A particular concern at present is that on most farms the potentials of plants and animals are far from realised. Average yields of many crops are no more than half those obtained by some good farmers and these in turn are less than can be obtained in experimental work. Increased research is needed to seek the causes of poor yields so that their effects can be reduced in practice. Integration of different branches of agricultural science in research programmes which involve crop and animal management studies should go far to achieve this aim.

Diseases and pests of crops and animals still cause large losses. Breeding crops that have durable resistance to diseases and pests will always be essential. The development of resistance to control chemicals in some pests will lead to better ways of integrated control involving both chemical and biological methods. More attention is needed to neo-natal mortality in animals and to enteric and respiratory diseases. We still need to know more about the incidence and epidemiology of both plant and animal diseases.

Improvements in yield must not be at the expense of quality. Cereal quality is particularly important in bread-making and malting. From animals we require lean meat of acceptable quality which does not deteriorate in handling.

Nitrogen is the key element for crop growth in many situations and is contained in all plant and animal proteins. Nitrogen fertilizers need much energy to make them, yet are essential in all intensive agriculture. More work is needed on conserving nitrogen in soil–plant–animal systems so that it can be used efficiently without the losses which cause pollution of water supplies by nitrate, or of the upper atmosphere by nitrogen oxides—both are causes of current concern to environmental scientists. An increased contribution by legumes to the nitrogen needed by pastures will diminish the need for fertilizers. Research now in progress at the ARC's Unit of Nitrogen Fixation may lead to new ways of fixing nitrogen from air by imitating in a chemical plant the activity of the enzyme nitrogenase (responsible for all biological systems of fixation); other, equally important, work is being done to endow non-leguminous plants with the ability to fix their own nitrogen from the air.

Finally, it is just as important to avoid waste as it is to produce more. Large losses of food now occur in harvesting, storage, distribution and use—these are estimated to be, altogether, a third of the total produced on farms. Research to avoid waste, and to make use of the wastes from crop and animal farming will be an essential part of our effort towards yet higher productivity.

ACKNOWLEDGMENTS

The authors acknowledge with gratitude the assistance they have received from colleagues at the Agricultural Research Council and from ARC Institute Directors in the preparation of this article.

SCIENCE AND THE DEVELOPMENT OF NUCLEAR ENERGY

R.M. LONGSTAFF

United Kingdom Atomic Energy Authority, London

POWER FROM FISSION

Introduction

This chapter is about the development of nuclear power in Britain, from the post-war era when 'atoms for peace' were coming into the public eye as a desirable follow-up from the successful weapons tests, to the present day when many of the technical problems have been solved. The status and future development of nuclear power now seems likely to depend more upon public confidence and acceptance than upon specific technological advances.

One obvious indication of growth is the burgeoning of nuclear electricity generation, which rose from zero in 1952 to 14% of the country's electric power needs in 1977. Another, and in some ways more significant sign, is the increase in the number of UK University Departments concerned with various aspects of nuclear science or engineering: in 1952 the 'World of Learning' yearbook did not list any, but in 1977 it named 16.

How Nuclear Power Stations Work

The term 'nuclear power' as used here means electricity produced by burning uranium in a 'reactor' at a power station, and using the heat to raise steam to drive conventional turbines and generators. Uranium burns in quite a different way from coal or oil and in doing so it liberates thousands of times as much energy per ton of fuel used. It is not a chemical process at all in the sense of being a combination of uranium with air or anything else: it is a breaking up of some of the uranium atoms themselves into two nearly equal parts and two or three much smaller fragments. The process is called 'nuclear fission' and the two halves (which are themselves atoms of intermediate weight chemicals) are called 'fission products'. The small fragments, sometimes two and sometimes three in number, are called 'neutrons'. Neutrons, with protons and electrons, are the constituent parts of atoms everywhere.

All these pieces fly apart at very great speeds, striking neighbouring atoms and making them move about rapidly, i.e. they get hot. It is this heat that is exploited in nuclear power generation. The neutrons either escape from the

69

system or are absorbed into other atoms. When one of the neutrons is absorbed into an atom of the right kind (or 'isotope') of uranium—uranium 235—it starts off the process of fission all over again, liberating yet more neutrons. In this way, fission is kept going as a nuclear chain reaction analogous to the chemical chain reaction of combustion; the nuclear reaction is not, however, sustained by heat but by the neutrons that it produces. As well as the energy of motion of the fragments, which shows itself as heat, fission gives off additional energy in the form of radiation, mostly gamma-rays which are very like X-rays but generally higher in energy. The level at which the reaction proceeds is controlled by inserting neutron-absorbing 'control rods' to absorb surplus neutrons. Because neutrons are easily detected and their population density is accurately and quickly measurable, very close control can be kept on reactor power output. Some of the neutrons produced in fission are absorbed by impurities, structural components, etc., and some get away from the reactor and are lost. For any given design of reactor, the smaller its physical size the greater will be its surface area in proportion to the mass of uranium in it, and hence the greater will be the neutron losses. A certain critical mass of uranium (which will differ for every reactor design and specification) is needed before the fission chain reaction can start.

Uranium 235 atoms occur in nature in the ratio of only about one in every 140 atoms of uranium, the other 139 atoms being uranium 238, which will not undergo fission in this way. However, neutrons stand a much improved chance of causing fission in uranium 235 if they have first been slowed down by repeated collisions with atoms of light elements such as hydrogen (in water) or carbon. The presence of substantial amounts of a 'moderator' in the fuel is, in fact, essential in a reactor unless the concentration of uranium 235 is substantially increased. This can only be done by a costly isotope separation process known as 'enrichment'. Nearly all of today's nuclear power stations use slow ('thermal') neutrons with only slight enrichment, if any, of the uranium.

Neutrons absorbed by uranium 238 are also lost to the chain reaction. However in some cases they form a heavier isotope of uranium, uranium 239. This is radioactive (i.e. its nuclei are unstable) and it changes in a matter of days to neptunium 239 which is also radioactive and changes in turn to plutonium 239. These last named elements do not occur in nature except in minute traces. Plutonium 239, like uranium 235, is fissionable but only part of it gets burned up in the reactor where it is formed. It is more useful if it is separated out and used as fuel in reactors specially designed for it—'fast reactors' which have no moderator—or it can be used in its original role as an explosive for nuclear weapons. Plutonium is also radioactive but it changes very slowly indeed, with a half-life of 24 000 years, i.e. it will take half of the atoms in any given quantity 24 000 years to make the change.

The State of the Art in 1952

At the start of the period covered by this book, peaceful nuclear power had grown to more than just a twinkle in the eyes of a few scientists—it was a hard gleam in the eyes of engineers. Weapons had been made and tested in the United States and Russia, and Britain was on the point of joining the 'club'. Reactor designs intended for plutonium production for weapons could now be reconsidered for the production of useful power. This meant reversing the priorities for the output of reactors: hitherto plutonium production had been their main object, while the heat produced from fission had been nothing but a costly nuisance to be dissipated somehow. Now it was the heat that was to be important, with plutonium as the by-product.

Britain was already operating a factory for the production of uranium reactor fuel at Springfields in Lancashire, and two large air-cooled plutonium-producing reactors at Windscale in Cumbria, with a plutonium separation plant alongside them. An isotope enrichment plant was under construction at Capenhurst in Cheshire. Engineering design and construction were concentrated at Risley in Lancashire under Christopher Hinton and Leonard Owen, while at Harwell on the Berkshire Downs the Atomic Energy Research Establishment was in full operation under the leadership of John Cockcroft. The factories and plants were in full production and Britain's first nuclear weapon was about to be tested. Some effort was becoming available for research into the peaceful applications of atomic energy and in particular the generation of electric power. At this time it became apparent that at least one more plutonium-producing reactor would have to be built for military needs. Why should not the heat that this would inevitably produce be used to raise steam for a power station of a commercial size, perhaps even at a commercial cost? So began the design study which came to be known under the acronym of PIPPA—Pressurized Pile for producing Power and Plutonium.

Opportunities and Constraints

Public opinion at this time associated atomic energy almost entirely with the toadstool clouds of the two atom bombs dropped on Japan, and the mysterious and dreadful effects of their radiations. Now, almost at the start of our period—on 2 October 1952—Britain's first atomic bomb was exploded and Dr. Penney, under whose leadership it had been made and tested, returned from Montebello, Australia, to become Sir William. The bomb became the highlight again and the public reaction was a mixture of pride and fear. To some it meant that all research in atomic energy should be shunned, while to others it meant that the way was now open to concentrate on the peaceful applications of this new force.

The research, development and production organization that had been set up to give Britain her atomic bomb and was now to provide her with nuclear power as well, was concerned with the full range of work from painstaking observations leading to basic understanding of scientific principles—the kind of work normally associated with university research departments—right through to the details of engineering structures and chemical manufacture. It was at the intermediate stage of project-orientated research that establishments such as Harwell, and the research and development laboratories of the production factories, had the task of organizing and carrying out highly specialized research involving quite novel equipment and techniques. This had to be developed largely from scratch and required great ingenuity and application, and a quite new kind of co-operation between scientists, who knew what they wanted to achieve and the kind of apparatus that they needed to do it, and the engineers and craftsmen who had to turn these ideas into hardware, often with no precedent to go on, no proper drawings and not much time.

Some of the Problems

The task now was to design and build a nuclear power station that would work, basing it on the experience gained with the reactors of the weapons programme.

If we now glance briefly at the reasons why one particular type of reactor was chosen for the British nuclear weapons programme, we shall understand better why Britain has followed lines of development different from those of most other countries, particularly USA. When it was decided, as far back as 1947, that Britain was to have nuclear weapons of her own, a choice had to be made between making them from highly enriched uranium 235 or from plutonium. To prepare pure uranium 235 requires very large quantities of electric power for the isotope separation plant and Britain was not in a position to provide this on the right time scale, and the engineering problems were formidable. It seemed, however, that it might be possible to build plutonium-producing reactors and associated chemical plant within a few years. The Americans, in contrast, had adopted both courses and as a result they had a very large isotope separation plant as well as plutonium-producing reactors. Britain had little choice but to build plutonium-producing reactors, fuelling them with natural (unenriched) uranium. Consideration of the physics of nuclear reactors (as well as war-time experience in North America) made it clear that this could indeed be done, but the reactors would have to be very large and would have to use heavy water, beryllium (or its oxide) or graphite as the moderator. The production of heavy water again involved the use of too much electricity as well as the development of a suitable separation technology, while beryllium was very expensive and little was known about its properties and manufacture except that it had difficult metallurgical characteristics and was very poisonous. This left graphite as the only choice, and it was a substance which British industry was already capable of manufacturing in large quantities and in grades of good commercial purity.

There was also a choice of coolant, the fluid to carry the heat from the fuel to the boilers. This could be either water as used in the American reactors, or a gas, for which air, carbon dioxide and helium were the most likely. If water was to be used it would be needed in very large quantities and at a very high degree of purity, otherwise problems would arise from the radioactivity induced in the dissolved impurities as the water passed through the reactor. It turned out that the only suitable body of water for such an installation was the River Morar in Western Scotland, but fortunately there was no need to incur the environmental sacrilege that its use would have meant: at about the time that the decision had to be made it became apparent that air cooling would after all be feasible, in spite of earlier doubts. It had just been established that deep fins could be fitted to the aluminium cans protecting the uranium fuel rods, which would improve heat transfer sufficiently for air cooling to be satisfactory, without their capturing too many neutrons.

The decision to build two big air-cooled reactors on a site near Seascale in Cumbria was taken in 1947. This was even before Britain's very first research reactor, GLEEP, had been commissioned at Harwell. GLEEP was a small air-cooled reactor based on natural uranium metal and a graphite moderator, and producing so little heat that it needed no coolant. It was followed three years later by the larger air-cooled reactor BEPO based on similar principles but producing 6000 kilowatts of heat. Both of these reactors were essentially tools for research, and were not intended as prototypes for the plutonium-producing reactors. Nevertheless, being based on the same principles, they were able to provide much useful information for the design and commissioning of the latter. Progress on the Cumbria site, which came to be known as Windscale, was rapid and just before the start of our period (in October 1951) the second of the two reactors was commissioned. Both reactors worked well until one of them was badly damaged by fire in 1957. This led to the closure of both reactors, but by that time dual-purpose reactors based on the PIPPA studies were in operation to fill the gap.

The PIPPA Design Study

Decisions had by now been taken on a number of points which fixed the parameters for the design and construction of the PIPPA reactor. The fuel was to be natural uranium metal in the form of rods just over an inch in diameter, protected by magnesium alloy tubes ('cans') fitted with close-set spiral fins to aid heat transfer. They would be stacked in vertical holes spaced seven inches apart in the graphite moderator (figure 1). Heat would be carried from the fuel elements to the boilers by compressed carbon dioxide gas, so the whole of the core containing over 100 tons of uranium and many thousand tons of graphite would have to be inside a vessel strong enough to withstand the gas pressures at which the system would operate. Much depended on the maximum thickness of steel plate which could be made and welded to the very high standards required. At this time it was considered that two inches was the maximum thickness which could be relied upon, and this would enable the vessel to contain the gas safely at a pressure of 100 lb per square inch (seven times atmospheric pressure). The pressure vessel would be in the form of a cylinder 20 metres high by 11 metres diameter with a domed top fitted with four large ducts to carry the coolant gas from the top of the reactor to the top of the boilers. Here it would pass down over

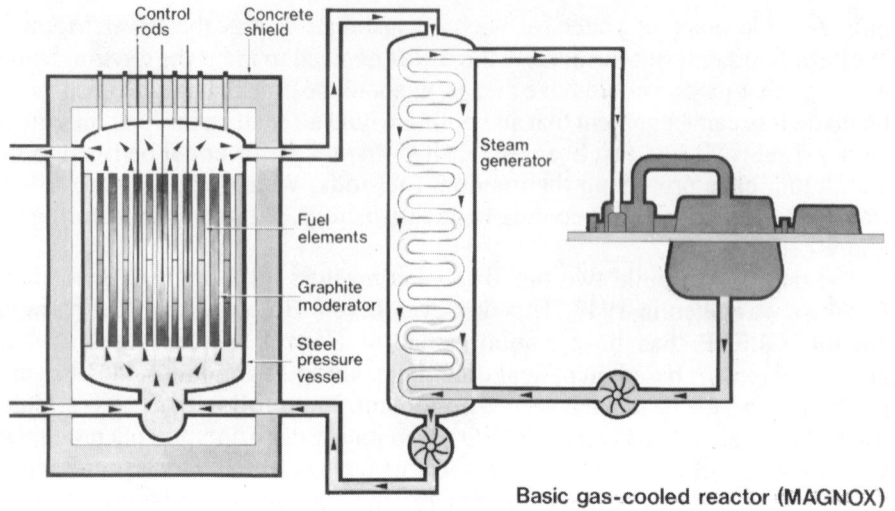

Basic gas-cooled reactor (MAGNOX)

FIGURE 1. Schematic diagram showing the essential elements of a gas-cooled nuclear power station: reactor, boiler and turbo-alternator.

a large number of water tubes, returning to the bottom of the reactor vessel via four powerful blowers.

The Fuel for PIPPA

Meanwhile one of the most important things to be settled was the design of the fuel elements. It is here that nuclear scientists and engineers had (and still have) to cope with problems which have few parallels in other industries. Consider what nuclear fuel has to do and to endure. Its job is to provide heat at a controlled rate, and with safety and economy, for the whole of the time that it remains in the reactor. This may amount to several years, and during this time it must maintain its mechanical strength and integrity, it must not corrode nor react with the coolant, and it must not allow the release into the coolant gas of any of the radioactive fission products that build up in it. These are some of the main reasons for enclosing the fuel in a can, but the latter also serves the additional purposes of enhancing the structural strength of the fuel element and, by close-set spiral fins, of improving the transfer of heat from it to the coolant. It must not, however, absorb an undue proportion of the neutrons and this imposes a major limitation on the choice of materials. The material chosen for PIPPA was an alloy of magnesium with small additions of aluminium, beryllium and copper. This was known as 'Magnox', a convenient name which soon became applied to the fuel itself and is now the generic name of the type of reactor in which it is used.

In operation the fuel has to withstand the most viciously difficult conditions. Every time that a uranium atom splits two new atoms appear in its place, each of them taking up not very much less room than the parent uranium atom. They immediately fly apart with great vigour, shaking up all the atoms around them, until they finally settle down some few atoms' breadths away from where they originated. Since nearly all solids, including uranium, are built as an orderly arrangement of atoms known as a 'crystal lattice' the sudden appearance of two new atoms near to where previously there was only one, will clearly cause considerable disruptions and strains in the structure of the solid. Some of the fission products are gases, and these tend to collect in bubbles along the crystal grain boundaries. Furthermore, two or three neutrons are produced in each fission and these also strike atoms near and far and displace them from their proper places. Together these causes of disturbance, combined with the elevated and varying temperatures at which the fuel has to operate to produce useful power, pose a severe strain on its metallurgical structure and physical integrity. The longer the fuel element is in the reactor and the higher the rate at which fission is taking place within it, the more rapidly will its structure become damaged. This 'irradiation damage' is one of the major limiting factors associated with nuclear fuel.

Another limiting factor is the mechanical damage caused to the fuel by repeated heating and cooling as the reactor is brought up to full power and shut down. This has the general effect of causing changes in the shapes of the crystal grains, leading to a roughening of the metal surface and a general distortion of the structure of the rods. Furthermore, the coefficient of expansion under heat of uranium metal is much lower than that of Magnox alloy so the can tends to expand much more on heating than does the fuel rod, but when the reactor is shut down it contracts rapidly on to the still-hot uranium rod which shrinks less than the can would. Repeated temperature changes have the effect of stretching the can beyond the end of the rod, even to the extent of rupturing it. This mechanical trouble, known as 'ratcheting', was finally overcome by machining numerous shallow grooves around the rod at intervals along its length and then pressurizing the comparatively soft Magnox can into the grooves so that it is mechanically bound to the rod and must perforce expand and contract with it.

A further way in which damage can take place is through the bending or bowing of the uranium rods under their own weight. This process, called 'creep', is aggravated by radiation and high operating temperatures. Metallurgical improvements in the fuel and canning material, cans designed for strength, and the fitting of external braces or struts, are all used to reduce the trouble to manageable levels.

In the PIPPA studies irradiation damage to the fuel itself was tackled initially on traditional metallurgical principles, by trying various likely additions to the uranium metal to see what improvement could be brought about by alloying it. Small additions of iron and aluminium were found to have an encouragingly beneficial effect, though the reasons for this were not at the time clear. Intensive research revealed that these added elements introduced discontinuities into the structure of the uranium crystals, resulting in stresses which tended to lock the constituent atoms into place and prevent internal movements or changes of shape. Further, they provided anchoring points for incipient bubbles of fission product gases, preventing these from coalescing and growing to sizes that would lead to swelling or distortion of the uranium. Changes due to thermal stresses and to creep were tackled empirically along similar lines. Later research revealed the precise mechanisms of these improvements and enabled them to be made even more effective, both in respect of PIPPA fuel and in more advanced fuels for later reactor designs.

Calder Hall

The scene was now set for the PIPPA design studies to crystallize into reality. In February 1953 the Government authorized the construction of a single PIPPA-type (dual-purpose) reactor on a site just across the Calder river from the Windscale works. Within a few months a second reactor was authorized, the two to comprise a single power station. In 1955 another twin-reactor power station of identical design, Calder Hall B, was ordered for the same site, and at the same time authorization was given for a virtually identical four-reactor station at Chapelcross in Annan, Scotland.

Calder Hall (figure 2, plate I) was officially opened by Her Majesty the Queen on 17 October 1956, when the first reactor started delivering electricity to the National Grid at the rate of some 30 000 kilowatts. Calder Hall and Chapelcross are now being operated solely for electricity production with each of the eight reactors capable of feeding 50 megawatts (50 000 kW) of electricity to the grid system. Design expectations have thus been exceeded and all eight reactors have run with very little trouble right from the start. Calder Hall A was the world's first full-scale nuclear power station. With its sister stations at Calder Hall B and Chapelcross it still holds many world records for performance: already more than one of the reactors has exceeded the design life of 20 years, and there are now plans to extend the useful life of Calder Hall into the 1990s. With hindsight it may perhaps be thought that the most remarkable achievement of all was the completion of the world's first industrial power reactors by the scheduled dates and within the allocated expenditures.

At the time that Calder Hall was being built, all atomic energy work came under the control of the Ministry of Supply (Department of Atomic Energy) which was of course part of the Civil Service. It was apparent that this was not the best arrangement and in 1954 the United Kingdom Atomic Energy Authority was formed as the statutory body responsible for the development of atomic energy.

Britain's First Nuclear Power Programme

That the Calder Hall design was likely to prove successful was apparent even before the first reactors came into operation. Accordingly in 1955 the

Government announced a ten-year programme of nuclear power under which improved versions of the Calder Hall reactors were to be built by British industry for the generating boards, to produce up to 2000 megawatts of nuclear power. This programme was later greatly expanded to an output of 5000 megawatts but over a time-scale three years longer.

Design work started at once on a succession of stations of essentially the same type as Calder Hall, but with each successive one designed to be larger and more efficient than its predecessors. The first of these, at Berkeley in Gloucestershire, came on power in 1962 with an electrical output of 276 megawatts, while the latest, at Wylfa in the Isle of Anglesey, came on power in 1971 and has an output of 1180 megawatts. See table 1.

TABLE 1

Britain's Nuclear Power Stations 1956–1977

Name and Location	Design Capacity MW (electrical)	Date on Power
Magnox Power Stations		
Calder Hall,* Cumbria	200	1956
Chapelcross,* Dumfries	200	1958
Berkeley, Gloucestershire	276	1962
Bradwell, Essex	300	1962
Hunterston 'A', Strathclyde	320	1964
Hinkley Point 'A', Somerset	500	1965
Trawsfynydd, Gwynedd	500	1965
Sizewell, Suffolk	580	1966
Oldbury-on-Severn, Bristol	600	1968
Wylfa, Gwynedd	1180	1971
Advanced Gas-cooled Reactor Power Stations		
Windscale,* Cumbria	33	1962
Hunterston 'B', Strathclyde	1320	1976
Hinkley Point 'B', Somerset	1320	1976
Dungeness 'B'	1200	(1980)
Hartlepool, Cleveland	1320	(1981)
Heysham	1320	(1981)
Fast Reactor Power Stations		
Dounreay Fast Reactor,* Highland	14	1959–1977
Prototype Fast Reactor,* Dounreay, Highland	250	1975
Steam-generating Heavy Water Reactor Power Station		
SGHWR, Winfrith, Dorset*	100	1967

The stations marked * were built by the UKAEA to prove the design. The remainder were built to the orders of the Electricity Generating Boards, who operate them commercially.

There were of course many improvements in design and in operating characteristics, most of which resulted from the application of the engineering lessons learned in the earlier reactors. Perhaps the most significant alteration was the use, at Oldbury and Wylfa, of prestressed concrete to form (in conjunction with a steel lining) the combined pressure vessel and radiation shield for the reactor core. This allowed the use of higher coolant gas pressures and therefore more effective transfer of heat from the fuel to the boilers, without the development of techniques for welding ever thicker sections of steel: the two

inches considered the maximum thickness at the time of Calder Hall had reached
to over four inches in some of the later stations.

Of course there were problems, not least those due to continually breaking
new ground rather than replicating a proven though imperfect design. But most
of these difficulties were overcome and the nine Magnox power stations are now
regarded by the Generating Boards as the reliable workhorses of the electricity
industry. Today they produce about 14 per cent of Britain's electricity, at a cost
significantly lower than that from coal-fired or oil-fired power stations built over
the same period.

Fast Reactors

At the time that Calder Hall was being designed and built, scientists and
engineers in Britain and abroad were already working on a more advanced
system which held out the possibility of very much more effective utilization of
uranium, the 'fast' reactor.

Studies of nuclear physics, and in particular of the details of how neutrons
moving at different speeds interact with nuclei of the various uranium and
plutonium isotopes, showed that it should be possible to design reactors without
any moderator and with a very small core containing a high proportion of
fissionable material (uranium 235 or plutonium) together with uranium 238.
These should show very good neutron economy and should indeed provide quite
a large surplus of neutrons over and above those needed to maintain the chain
reaction. Some of these could be trapped in a surrounding 'blanket' of uranium
238, which would then progressively become converted to plutonium. This
opened up the possibility, later confirmed by experiment, of 'breeding' more
fissionable material in the blanket than the core itself consumed—in other words
a reactor that could if need be produce more fuel than it used. A supply of
plutonium fuel might be built up in the course of time sufficient to fuel other
fast reactors. To check this line of reasoning and to test the accuracy of the
predictions, a small fast reactor 'Zephyr' was built at Harwell in 1953. It had a
core about the size of a $2\frac{1}{2}$ litre tin of paint, consisting of pencil-sized rods of
plutonium sealed in metal tubes, together with some rods of uranium metal. This
was surrounded by a blanket of natural uranium rods which served both to deflect
some of the surplus neutrons back into the core and to provide for the breeding of
plutonium. The system was designed to have a very low power output so that
cooling would be unnecessary (the Z in Zephyr stands for 'zero energy').
Nevertheless elaborate safety precautions were taken in its design and operation.

Zephyr confirmed the feasibility of the fast reactor system and was supported
by parallel work in the United States and Russia. By 1952 preparatory design
work was in hand for a fast reactor large enough to drive a small power station,
and in 1954 authorization was given for building a fast reactor power station with
15 megawatt net electrical output at Dounreay on the north-east tip of Scotland.
'DFR' (Dounreay Fast Reactor) started operating in 1959 and by 1960 was

feeding power into the north of Scotland grid. Initially it was fuelled with highly enriched uranium (75% U 235).

Because the core of a fast reactor is so small by comparison with one of similar heat output in, for example, the Magnox design the rate at which heat has to be taken away to the boilers is phenomenally high. The problem of its removal did not arise in Zephyr, but in DFR some 60 megawatts of heat had to be removed from a core rather smaller than an ordinary domestic dustbin. Gas cooling at the current stage of technology would not have been capable of doing this, and water cooling would introduce moderating material which would nullify the advantages of the fast reactor. Of the coolant materials considered at the design stage the most likely ones were liquid sodium and sodium–potassium alloy. The latter remains liquid at ordinary temperatures and was chosen for DFR but it was expected that liquid sodium alone would be used in later fast reactors.

Sodium as a Coolant

As sodium reacts violently with water it is vital that it is kept firmly out of contact with the water or steam in the boilers. Accordingly an indirect heat transfer system was devised whereby sodium from the core circulates through a primary heat exchanger system of 24 separate units, where it gives up its heat to a second, entirely separate, circuit of sodium which goes to the boilers to raise steam for the turbines. Very special attention had to be paid to the quality of all the pipework and welds throughout the sodium system and the experience gained in the construction of DFR contributed greatly to the technology and quality-control of welding stainless steel. Although working with liquid sodium on this scale was unprecedented in Britain, no major difficulties were encountered.

One of the advantages of liquid sodium is that it does not boil at the core temperatures (about 350°C) of the reactor so, unlike water or compressed gas, it does not need to be contained in a pressure vessel. The core of the reactor and its blanket of uranium 238 are bathed in liquid metal that flows rapidly upwards past the fuel elements and through the primary heat exchangers, from which it is pumped back to the reactor tank. Because sodium is a metal it conducts electricity very well and, by passing a current through it, it can be made to move in a magnetic field. It is thus possible to pump the liquid metal by electromagnetic pumps that have no moving parts whatever, and this was done in parts of the DFR system.

The fuel for DFR consisted of uranium metal enriched to 75% uranium 235. It was canned in niobium, but it was anticipated that stainless steel would be used in commercial reactor fuels. Similarly the enriched uranium metal would be replaced by fuel consisting essentially of a mixed oxide of about 20% plutonium and 80% uranium 238. DFR was designed and engineered with safety as a major consideration, but at the same time the 135-foot steel sphere that forms such a prominent landmark at Dounreay provides an additional safeguard designed to contain any radioactive material that might be released in the event of an accident in the reactor, and to withstand the changes of pressure that would result if the liquid sodium caught fire.

Research in DFR

Initially DFR's main function was to test the physics of the fast reactor system, studying in particular such things as the rates of fuel consumption and of the breeding of plutonium, and the mutual effects that changes in temperature and rate of power output have on one another and on the control and operating characteristics of the reactor, under a wide range of conditions. Later it took on many of the functions of a test reactor for studying engineering and materials problems and heat transfer.

A major group of experiments was concerned with studying the performance of fast reactor fuels. Advanced post-irradiation examination (PIE) facilities were installed to make it possible to carry out detailed examination of the mechanical, physical and metallurgical conditions of the highly radioactive irradiated fuels, as well as following the nuclear and chemical changes taking place in them. DFR was also used to develop the plutonium-bearing fuels used in its successor at Dounreay, the Prototype Fast Reactor (PFR) which was to be the next step towards the economic development of fast reactor systems.

DFR approached the close of its useful life soon after the much larger Prototype Fast Reactor—which can do all that DFR did, and far more—started operating in 1974. DFR was finally closed down in March 1976 after some 17 years of highly successful operation.

Prototype Fast Reactor (PFR)

PFR (figure 3, plate II) was designed as the smallest reactor from which the information necessary for the design of fully commercial fast reactor power stations could be obtained with confidence. It also supplies steam to the generating unit of a power station having a net design output of 250 megawatts. Although much smaller than any commercial fast reactor will be, PFR has been designed to gain experience applicable to full sized plant having an electrical output of some 1300 megawatts. For instance, the boilers, the sodium pumps and the core and blanket fuel elements are the same size as those that will be used in a commercial fast reactor; the major difference is in the number of units rather than their individual size, and no component units in the commercial reactor will be more than double the size of those in PFR. Its core fuel charge is made from mixed oxides of plutonium (about 20%) and uranium. PFR achieved full thermal power output in 1977, when it was feeding 200 megawatts of electricity to the grid.

The Advanced Gas-cooled Reactor (AGR)

The Magnox series was clearly not the ultimate in graphite-moderated gas-cooled thermal reactors. The use of uranium metal fuel in Magnox alloy cans imposed upper limits on the temperature of operation, and therefore on the thermodynamic efficiency of the steam turbine system. Further, the endurance of the fuel under prolonged irradiation and repeated thermal cycling was limited, in spite of all that the metallurgists could do. The maximum 'burn-up' that could be effectively guaranteed was about 4000 megawatt days per ton—that is to say a ton of uranium would enable the power station to produce the equivalent of 4000 megawatt days of electricity (96 million kilowatt-hour units). It is worth noting that at about the time that Calder Hall was opened, there were serious doubts about achieving even half of this output with certainty. Moreover, the Magnox stations were characterized by their large physical size and therefore high capital cost per kilowatt of installed power. To reduce this the heat would have to be got out of the fuel at a higher rate, and this again meant both higher working temperatures for the fuel and a higher surface area in relation to its mass, i.e. thinner fuel rods.

In August 1957, less than a year after Calder Hall started up and more than four years before the first of the big Magnox stations came into operation, design and development work started on the Windscale Advanced Gas-cooled Reactor (WAGR). This was to be a prototype on which, if successful, a second programme of gas-cooled nuclear power stations might be based. The essential advances were in the fuel, which was to be uranium oxide pellets enriched to about 3% U 235 and canned in stainless steel tubes each about half an inch in diameter; they would be grouped in clusters. The coolant as before was to be carbon dioxide, at the comparatively high pressure of about 19 atmospheres. The new fuel permitted a gas outlet temperature of up to 575°C (the highest Magnox gas temperature was 410°C). This would allow steam temperatures of 454°C, over 50° higher than in the hottest Magnox stations.

WAGR came into operation in 1963. It was intended to generate 27 megawatts of electricity but in fact it was eventually operated at a steady 33 megawatts. As a test-bed for the AGR system and the oxide fuels to be used with it WAGR was highly successful. It was too small, however, to provide a satisfactory basis on which to design the very much larger commercial AGRs that were to follow it. Although the nuclear aspects of the scaling-up could be handled successfully, various engineering problems arose which could not reasonably have been foreseen from the operation of the small prototype.

Britain's Second (AGR) Nuclear Power Programme

In 1964 the Government authorized Britain's second nuclear power programme, based on the AGR system and comprising five twin-reactor power stations each with an electrical output of around 1300 megawatts. In the light of knowledge at the time, these were expected to generate electricity more cheaply than the Magnox stations (which even then were generating at costs not far different from coal-fired stations). Two AGR's, at Hinkley Point (figure 4, plate III) and Hunterston, came into operation early in 1976, fulfilling all that was expected of them. Their technical success seems assured and when all five of them are in operation, along with the Magnox stations, about 20% of Britain's electricity will be generated by nuclear power.

High Temperature Gas-cooled Reactors

The Advanced Gas-cooled Reactors, though a great improvement on their predecessors, the Magnox Reactors, do not represent the limit of development for the gas-cooled graphite-moderated reactor concept. Work at Harwell as early as 1956 had suggested that it should be possible to design a reactor in which the fuel was not, as hitherto, in the form of substantial pieces of uranium (or its oxide) distributed at fixed places through a stack of graphite blocks, but would take the form of small particles distributed more or less homogeneously through the graphite moderator. Such a reactor could operate at very much higher temperatures than hitherto, of the order of 1200°C. Further, it opened up the possibility of using thorium as a 'fertile' material in which fissionable uranium 233 would be produced, analogously to the production of fissionable plutonium in uranium 238. By 1959, a tentative design for a high temperature gas-cooled reactor had been prepared and in the same year an agreement between the countries of the Organisation for European Economic Co-operation—Britain, the six EURATOM countries and Austria, Denmark, Norway, Sweden and Switzerland—was signed initiating the 'DRAGON' project. The project was located at the newly opened Atomic Energy Establishment at Winfrith in Dorset, and was truly international in character, with working teams of mixed nationalities. Construction started on the reactor in 1960 and it came into use at full power in 1966. It was designed to test the feasibility of the high temperature gas-cooled reactor system, to develop and test a range of experimental fuels (including breeder fuels based on the thorium cycle), to develop the use of helium gas as a coolant, to study ways of preventing (and cleaning up) fission gas releases into the coolant, and to reduce corrosion of the graphite. Finally it would provide operational experience with this type of reactor.

A major potential application of the high-temperature gas-cooled reactor system was for producing process heat for industry, for example for steel making. However, the industrial interest shown was not sufficient to justify extensive work in the UK on this aspect, although it was discussed within the steel industry both here and overseas. The DRAGON project came to an end in March 1976. It had provided a convincing demonstration of the feasibility and advantages, and some of the problems, not only of the system and its variations but also of continued co-operation between scientists and engineers from different countries, each of whom brought with him his own specialized discipline, experience and background to a single major project.

Water Reactors

Britain's commitment to the gas-cooled thermal reactor concept was not total: from 1957 onwards some effort was being put into developing a system in which water was the coolant. Already in the USA virtually all the power producing reactors under construction or on order were of the pressurized-water or boiling-water type, in which ordinary water was both moderator and coolant. Canada too was concentrating on water-cooled reactors, but of a different kind. She developed the 'CANDU' (Canada deuterium uranium) concept, where heavy water (deuterium oxide) acts as both coolant and moderator, enabling natural (unenriched) uranium oxide canned in zirconium alloy to be used as fuel. None of these three North American systems were ideal for Britain, and a study was made of possible alternative designs. Eventually a pressure-tube reactor with a tank for the heavy water moderator was decided upon. The coolant, ordinary water, would flow upwards through vertical pressure-tubes fixed in the tank and would be allowed to boil as it rose past the fuel elements in the tubes. The steam would be collected in a steam-drum at the top, where it would be separated from entrained water and fed directly to the turbines. The uranium oxide fuel would have to be slightly enriched and canned in zirconium alloy.

In 1963, site work started at Winfrith on the prototype steam generating heavy water reactor (SGHWR) which came up to full power on schedule in January 1968, feeding 100 megawatts to the grid.

Britain's Third Nuclear Power Programme

SGHWR came fully up to expectations, proving reliable and easy to run. When Britain considered embarking on her third nuclear power programme, SGHWR was one of the contenders. The others were improved Magnox, AGR, HTR, PWR, BWR and CANDU. In July 1974, after considerable airing of views in the Press and lengthy debates in Parliament, the Government of the day came down in favour of adopting SGHWR for an initial programme of not more than 4000 megawatts over the following four years with (it was hoped) more to follow when confidence was established and demand stabilized. Since SGHWR was selected the position has changed substantially. Some reservations on the safety of PWR's have been cleared up, and changes in the economic situation now make it unlikely that a large programme of nuclear power station building will be undertaken in Britain in the near future. In addition, the first of the Advanced Gas-cooled Reactors are now working well and giving greatly increased confidence in that system. For these and other reasons, some technical, some commercial, it would be hard now to justify development of

SGHWR to commercial viability either for home use or for exports. Accordingly the Government is reconsidering the question of reactor choice and a fresh decision is likely to be announced soon.†

Fuel Manufacture

So far we have looked at nuclear reactor systems mainly in their capacity as power units for electricity generation, and at the demands that they make on the designers of the fuel that they use. Let us now look at some of the problems that arise in the chemical stages of the production, use, and reprocessing of fuel—the 'fuel cycle'.

Although Britain boasts no presently economic uranium deposits in her own territory (most of what we use is imported from Africa, North America and Australia), nevertheless teams from the Institute of Geological Sciences in London and from Harwell have played an important if indirect part in developing world resources. They have designed, and British firms are marketing, a range of portable instruments that can very effectively detect, identify and assay radioactive materials on or below the surface of the earth, or on the sea bed. Other scientists have studied novel techniques for uranium extraction, such as strong acid leaching from the ore *in situ* or the harnessing of bacterial action for the same purpose. Simultaneously, techniques have been worked out for the extraction from sea water of the small amount of uranium that is dissolved in it, should such expensive measures ever become necessary.

Most of the uranium imported into the United Kingdom for the nuclear power programmes has been delivered to the Springfields works of British Nuclear Fuels Limited (the state-owned company responsible for manufacturing and reprocessing nuclear fuel) in the form of concentrates or 'yellow-cake', which is essentially a crude oxide containing upwards of 50% of uranium. To be used as nuclear fuel the uranium has to be very pure indeed, and great improvements have been made in the purification techniques used at Springfields. Two major innovations were developed, right from the laboratory bench to full commercial scale. First came the multi-unit mixer-settler for continuous extraction of uranyl nitrate into a solvent and back in a very high state of purity into water. This represented an improvement not only over the existing batch process at Springfields, but also over the continuous extraction process in columns as developed earlier for the separation of plutonium at Windscale. The second improvement was the continuous multi-stage 'fluidized-bed' plant, in which the uranyl nitrate solution is first spray-dried and decomposed by heat to oxide powder, then after reduction with hydrogen to a lower oxide, converted to tetrafluoride by reaction with dry hydrogen fluoride gas. At all these stages the uranium compound is in powder form with the appropriate hot gas passing up

---

†In January 1978 the Government authorized the building of two new AGR power stations, and endorsed the nuclear industry's proposal for a design study of a PWR. At the same time it announced discontinuation of work on SGHWR.

through it, separating the grains and causing the mass of powder to flow from vessel to vessel like a liquid. Fluidized-bed techniques are now finding many applications in industry, the power-station furnace burning pulverized coal in a fluidized bed being an example currently under development in many parts of the world.

The only remaining batch process on the chemical side at Springfields now is the reduction of tetrafluoride to metal for Magnox fuel, by a thermal reaction with magnesium metal.

Improvements have also been made in the preparation of uranium hexafluoride to be fed to the enrichment plant at Capenhurst. The first stage, at Springfields, is the conversion of uranium tetrafluoride to hexafluoride. This was originally done in a batch plant where the tetrafluoride was made to react with chlorine trifluoride. In the new continuous process plant, fluorine gas is prepared by the electrolysis of magnesium fluoride (the slag that forms as a by-product of the reduction of uranium tetrafluoride to metal), and this is combined with the tetrafluoride in another fluid bed plant. Reconversion to oxide of the uranium hexafluoride after isotopic enrichment is also done in a newly developed integrated dry process which replaces the earlier wet batch process; it leads directly to uranium dioxide powder in a form suitable for fuel manufacture. These continuous process plants are run almost entirely by advanced automatic control techniques using sensors, digital computers and actuators, with human operators in supervisory capacities rather than in direct control. It has involved considerable advances in the science and technology of cybernetics.

In all the Springfields processes the fullest possible recovery and re-cycling of materials is practised. Not only are the acids, solvents and slags recovered and re-used but all effluents are thoroughly cleaned up and monitored, particularly for traces of radioactive or toxic chemicals.

These new chemical processes and plants at Springfields are founded on thorough understanding of the underlying principles of chemistry, physics and engineering. They represent in their design, installation and operation a very high degree of inter-disciplinary co-operation, which has also extended from the shop floor to top management and from the research laboratory to the plant manufacturer.

Enrichment

The isotope separation plant at Capenhurst (originally built to produce very highly enriched uranium for weapons) has also seen major improvements over the period. Initially the isotope separation process depended entirely on the gaseous diffusion of uranium hexafluoride through a porous membrane, with the lighter uranium 235 diffusing slightly faster. Thousands of units, connected in series and in parallel, are needed for effective separation, and each has its own compressor and membrane, together with forward and backward feeds connecting it with its neighbours. The process is expensive in electricity and capital cost, the layout is virtually fixed and there is not much flexibility in operation. Nevertheless it has been adapted to produce all the low-enriched materials required for the power programme.

The major advance, however, has been the design, installation and successful start-up early in 1977 of a quite different separation process using high-speed gas centrifuges instead of compressors and diffusion membranes. This has involved intensive scientific and engineering research and experimentation undertaken on an international basis between Britain (acting through British Nuclear Fuels Ltd), West Germany and Holland. Together these countries have formed two international companies, CENTEC to develop and manufacture the centrifuges themselves, and URENCO, to operate the process and supply low-enriched uranium hexafluoride to its customers, who include British Nuclear Fuels Ltd. The URENCO plant at Capenhurst is the first commercial-size centrifuge plant in the world, and has a throughput of 200 tonnes of separative work per year.

Reprocessing

Magnox reactors, AGRs and indeed any current thermal reactors, burn only a fraction of the uranium from which the fuel is made—about one third to one half of the fissile uranium 235 and none of the non-fissile uranium 238. Some of the latter is converted to fissile plutonium, of which however only a small fraction is burned. As we saw, the fuel elements have to be removed from the reactor towards the end of their useful life, while they still retain their physical integrity and before they start being net users of neutrons rather than producers.

In the days before nuclear power, fuel from the military plutonium-producing reactors at Windscale was taken out as soon as a worthwhile amount of plutonium had been generated and before further neutron capture caused its bomb-making quality to deteriorate due to the build-up of higher plutonium isotopes. It was then put through the nearby chemical reprocessing plant to recover the plutonium. In power reactors the aim is rather to get maximum energy from the fuel while it is in the reactor. To get plutonium out, at whatever stage it has reached, and to recover the residual uranium, the used fuel has to be dissolved in acid and put through a chemical separation plant of considerable complexity, using solvent extraction processes allied to those used in uranium purification.

Because of the radioactivity of the fission products and the toxicity of the plutonium, most of the processes take place under remote control, behind heavy shielding or in areas otherwise isolated from human contact. The original plant at Windscale was built on the assumption that maintenance in the ordinary sense of the word would be extremely difficult, if not impossible. In spite of this, however, it has been found practicable to do a surprising amount of successful maintenance and even development work in active areas.

Plutonium poses the further problem of 'criticality', which it shares with highly enriched uranium. This is the possibility of inadvertently setting off a fission chain-reaction simply by getting too much fissile material together in one place. Plant must be designed, operated and controlled in such a way that this is virtually impossible.

Magnox fuel reprocessing went ahead at Windscale in the original plant, and later in a new plant based on horizontal mixer-settlers rather like those at Springfields. A proposed new oxide reprocessing plant was the subject of a major public enquiry in 1977.†

Nuclear Wastes

Radioactive fission products, which comprise the major wastes from a reprocessing plant, arise as an acid liquid which is at present concentrated and stored in large stainless steel tanks, double-walled and monitored for leakage, cooled and stirred to remove the gradually decreasing heat of radioactive decay, and shielded by several feet of concrete. So far all the highly active waste from Britain's nuclear power programmes has been stored safely in this way, and although the method is unlikely ever to give trouble, it would clearly be simpler to store the waste as a solid, in which form it could not leak and would need less surveillance. Scientists at Windscale and Harwell are co-operating in bringing up to full industrial scale a process (already proven on the pilot scale) for converting the liquid wastes into blocks of solid glass clad in stainless steel. These will be very resistant to the action of water and will retain their integrity for a very long time indeed. Simultaneously work is in hand, on a national and international basis, to identify the kinds of stable geological formations in which these blocks can be safely buried out of Man's environment for ever. As part of a Common Market research programme, British geologists and waste disposal experts are studying granite and other hard rock formations, and also clay. They are studying too the possibility of burial below the deep sea bed.

Research Reactors

We have seen some of the scientific and technological advances made during the period in the mainstream of nuclear power development. Let us now examine some aspects of the scientific research that have provided support for these advances, and at some of the facilities and equipment used. The nuclear reactors GLEEP and BEPO have already been mentioned. The importance of research reactors can hardly be over-emphasized. Basically their job is to provide an abundant source of neutrons, and the environment in which to do the kind of experiments that need them, rather as a Bunsen burner provides a source of heat

†In January 1978 the 'Parker Enquiry' recommended that, subject to certain conditions, the plant should be built.

at the laboratory bench. The functions of research reactors in the context of nuclear power can be seen as three-fold. First they are used to throw light on the basic facts and figures of neutron behaviour and interactions. Secondly, they provide, sometimes in a much enhanced degree, the conditions of neutron bombardment that the materials and components of real power reactors will have to stand up to. Thirdly, they provide, on a much reduced scale of size, power output and radioactivity, the means of studying experimentally the physics of particular reactor systems.

Experiments can be done and measurements made either inside or outside the reactors. Inside there is a hailstorm of neutrons of energies covering the whole spectrum between those fresh from fission and those reduced to thermal equilibrium with the moderator atoms. If the reactor is designed and operated so that the energy distribution of the neutrons is the same as that in a power reactor, but the total 'flux' (number of neutrons per second passing through a given hypothetical target area in the reactor) is higher, performance tests can be carried out on power reactor materials and components in a correspondingly shorter time and under strictly controlled chemical, physical and mechanical conditions. Another way is to open a hole in the reactor shielding through which to let out a stream of neutrons which impinges on the experimental target. In this case it is possible to sort out neutrons of different energies by passing them through rotating 'neutron choppers' and time-of-flight tubes. These act as a kind of revolving door system, letting through batches of neutrons having the precise energies required for the experiment. In this way, very detailed information can be obtained on neutron interactions with target materials of all kinds. Even after thirty years of collaborative measurements made in laboratories all over the world, new information and more precise measurements are still coming in. For example, if a particular fission product, or a daughter product of its radioactive decay, has an unusual avidity for neutrons of a particular energy within the range of a reactor's operation, then its gradual build-up in the fuel may have an increasingly adverse effect on the reactor's performance.

Finely collimated neutron beams are used to obtain detailed information on the structure of solids by studying the directions in which the neutrons are scattered on striking the target specimen at different angles to its crystal axes.

We have likened materials-testing reactors to the laboratory Bunsen burner. Just as we may need to use a test tube to contain a chemical while we study its behaviour on heating, so we need some means of holding our nuclear specimen in the reactor where the neutrons are, in the physical and chemical environment that it will have to stand up to in use. This may involve providing the means of achieving, holding and monitoring temperatures, pressures and chemical surroundings, which may be very different indeed from those that obtain in the rest of the reactor core. In DIDO at Harwell there is a neutron flux thousands of times greater than that in any thermal power reactor, but the temperature in the core is only about 60°C and the pressure is not much above atmospheric. If then we have to test the fuel for a high-temperature reactor, we will need an environment of high-pressure helium at a temperature that may be well over

FIGURE 2. Calder Hall, the world's first full scale nuclear power station, was opened by HM the Queen on 17 October 1956. The four Magnox reactors (of which two are shown, each with four external boilers) together supply power to generate 200 MW electricity for the grid. Calder Hall was built by the UK Atomic Energy Authority and is operated by British Nuclear Fuels Limited.

PLATE II (LONGSTAFF)

FIGURE 3.   The Prototype Fast Reactor at Dounreay, built by the UKAEA to supply 250 MW of electricity to the grid, came on power in 1975. In the background is the spherical containment vessel of the experimental Dounreay Fast Reactor which came on power in 1959 and operated successfully until closed down in 1977.

FIGURE 4. Hinkley Point 'B' nuclear power station in Somerset came on power in 1976. It has two Advanced Gas-cooled Reactors designed to power turbo-alternators supplying 1320 MW of electricity to the grid.

PLATE IV (LONGSTAFF)

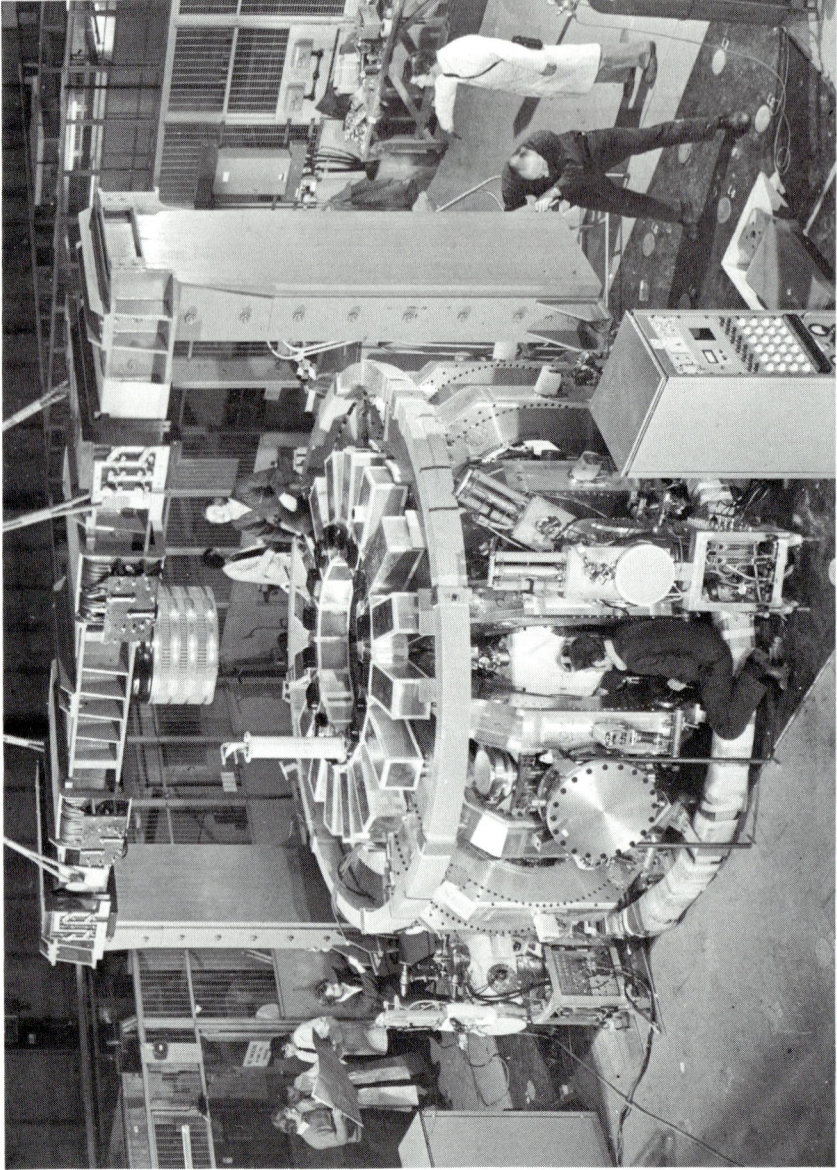

FIGURE 7. The DITE tokamak during final stages of construction. The upper half of the large transformer core is being lowered into place. The toroidal magnetic field coils (liquid nitrogen cooled) can be seen and also two large horizontal coils forming part of the transformer primary winding.

FIGURE 8.   A ring of plasma contained in the TORSO stellarator. In this stellarator the toroidal magnetic field coils and the helical winding are combined into one helical winding which can be seen supported outside the luminous plasma ring. Note that the plasma is well separated from the windings.

1000°C. The design and construction of rigs for doing this can take years and cost hundreds of thousands of pounds. Means have also to be devised of getting the specimen, and the rig itself, in and out through the thick shielding that surrounds the reactor, without letting out a dangerous stream of neutrons and other radiation.

Post-Irradiation Examinations

After suitable times of exposure, the test specimens have to be removed for post-irradiation examination under safe conditions using shielded 'caves' or cells and all the paraphernalia of remote manipulation. Adequate post-irradiation examination facilities, with the experience to operate them successfully, are essential to nuclear research and development at all levels. With the associated reactors they form a useful basis for collaborative research or contract services, and Britain is among the world's leaders in this field.

Accelerators

Charged-particle accelerators—the machines that used to be called 'atom-smashers' in the days when atoms were still popularly supposed to live up to the derivation of their name and be unsplittable—have for decades been valuable tools at all levels of nuclear research. One of the first major pieces of equipment at Harwell was the Synchrocyclotron which was completed in 1949 and which, after twenty-eight years and many modifications, is still operational. Now it is only one of a number of accelerators producing beams of charged particles that range from electrons and protons to the nuclei of the heaviest elements, and that reach energies corresponding to many millions of volts. Some of these provide continuous beams of particles, and some provide them in rapidly repeated pulses or in bundles of only a few nanoseconds duration (the time it takes for light to travel a few feet). Others provide beams having very precisely controlled energies. Beams of neutrons (which are electrically neutral and cannot themselves be accelerated) can be produced indirectly by making beams of charged particles of suitable energy impinge upon targets of selected elements. For example, deuterium beams directed on to a tritium target produce neutrons of energy corresponding to an accelerating potential of 14 million volts.

These accelerators, by reason of their precisely controllable output, supplement and extend the measurement work of nuclear reactors. They are also used to simulate in a relatively very short time—hours in some cases—the effects that years of exposure in reactors can have on materials: for example, the damage caused by many years of fast neutron bombardment on reactor-vessel steels can be simulated in a few hours. Chemical reactions can be studied under the influence of controlled radiation: for example, the important radiation-induced reactions between reactor coolant gas mixtures and the graphite moderator can be greatly speeded up. An advantage of using charged particles rather than neutrons from a reactor is that much less radioactivity is induced in the specimen under test, thus greatly simplifying its examination afterwards.

A major use of accelerators is as tools for fundamental research in physics. For example, particle beams act as precision probes for studying the structure and dynamics of solids at the atomic level. Also because of the precision with which the particle energies can be controlled, accelerators continue to be invaluable in measuring the detailed data of interactions between neutrons and nuclei. They also enable the fission process itself to be studied in detail.

When a beam of particles impinges on a target material, secondary radiations may be excited in the target which are quite characteristic of the particular elements composing it. This opens up the possibility of using particle beams as advanced analytical tools, and indeed this is one of their increasingly important uses. A wide range of such analytical techniques is now available, some of them of great value to the nuclear scientist. One example is the detailed surface examination of irradiated fuels and other reactor materials.

Beams of neutrons produced with the aid of an accelerator (or a nuclear reactor) can be used analogously to X-rays for producing radiographs (shadow-pictures) of the internal structure of an object. For this, a special film system is required which is insensitive to ordinary radiations (e.g. from radioactive materials) but which, unlike ordinary film, is sensitive to neutrons. Neutron radiography was first developed for the radiographic examination of highly radioactive used fuel elements and the like. It is now a recognized tool of industrial non-destructive testing, in which it complements conventional X-radiography by penetrating heavy materials like steel to reveal details of light materials (plastics, grease, water, etc.) that X-rays ignore.

Radiation Measurements

Many of the measurements that have to be made in the nuclear industry and in its research and development stages are measurements of 'radiation' in all its many forms. Because many kinds of radiation are (or can be treated as) streams of individual particles of sub-atomic size, they lend themselves to being counted as a succession of single events. The quality of a beam of radiation, whether emitted from a radioactive substance, from an accelerator (or the target of its beam) or from the core of a nuclear reactor, is determined by the nature of its particles or radiations, their energies and the number of them passing through a given target area every second. Although none of these radiations are directly apparent to the human senses, there are many ways in which they can be detected. Individual particles—electrons, ions, neutrons, gamma-ray photons, etc.—can be sensed as they reach a detector and their numbers added up electronically and displayed as a number of 'counts' taken over a given time. Or the general level of radiation intensity at the detector can be measured electrically and displayed as what is in effect the rate of energy deposition or 'dose-rate' at that point. Radiations can also be detected and measured by their effects on photographic film, as in the well-known film badges worn by all radiation workers, and by a variety of other techniques.

It is possible to separate the different kinds and energies of radiations or particles, and to measure only the relevant ones, rather as an optical spectrometer will separate the coloured light in any particular part of the visible spectrum and measure that. The nature and intensity of radiations at any point can be measured with great discrimination and accuracy and, because individual sub-atomic particles are being detected one at a time, with extreme sensitivity. The results of the measurements can be processed electronically for handling or display either digitally as a number, or in analogue form as readings on a dial or a curve on a graph.

Electronics

In the early part of the period reviewed, virtually all electronic equipment was based on thermionic valves. Pulse-counting and spectrometric equipment was bulky, it used a lot of current and therefore generated a lot of heat which had to be removed, it was hamperingly slow in its response, and it was far from reliable, especially at start-up. With the advent of the transistor and other benefits of 'solid-state' electronics, the position changed rapidly for the better, and thanks to the efforts of physicists and electronic engineers, many of them in this country, equipment became more compact, cheaper to make and to run, more versatile, more sensitive, much quicker to respond and, above all, far more reliable. In particular, components became smaller and designs moved through miniaturization towards micro-miniaturization, while integrated circuit blocks of modular construction were devised and built to international standards that applied both to the hardware itself and to the characteristics of the electrical signals used and produced by it.

Computers

In no field of electronics were such advances made during the period as in computers. There was a high degree of interdependence between the progress of nuclear energy and the development of computers, particularly of the large digital computers capable of receiving and processing information at very high speeds, and of accepting it from a number of sources simultaneously.

In the early days the designers of nuclear reactors were faced with stupendous tasks of calculating reactor dimensions and layouts of fuel, moderator, controls, structural materials, etc., and the dynamic effects of changes in temperature, coolant gas pressure, control rod movements, fuel burn-up etc. on the operating characteristics and safety of the reactor. Radiation levels due to fission itself and to the presence of fission products and of neutron-induced radioactivity, all had to be calculated over the whole operating life of the reactor.

These and countless other problems of no less difficulty had to be tackled—and solved—without the aid of any but what now seems to be the most rudimentary of calculating machinery. It was not until the late 1950's that computers were developed that were able to accept and handle data and present results at the kinds of speeds that the designers needed to allow optimization of design within a realistic time-scale. Faster and increasingly powerful computers were developed in and for the nuclear industry, especially as solid-state electronics succeeded the thermionic valve, and as micro-circuitry was developed. Now not only do the design calculations depend on the computer and its staff, but so does the practical control of the reactors themselves, and indeed of complete power stations and their integration with the national electricity distribution system.

Theoretical Physics

A large part of the initial stimulus for developing computers and computer programmes came from the needs of theoretical physicists and mathematicians working in regions that underlie the development of atomic energy. These include fundamental studies and calculations on nuclear structure, nuclear reaction rates and mechanisms, collision processes between two or more particles, the electronic structure of solids, the nature and effects of crystal lattice imperfections, etc. Examples of work on the engineering side (and therefore closer to practical problems) include theoretical studies on the fracture of solids, 'creep' under irradiation, the formation of corrosion films on metals, and radiation shielding calculations.

Heat Transfer and Fluid Flow

A major part of the design and structure of a nuclear power station concerns the transfer of heat, by means of a coolant fluid, from the fuel where it is produced to the boilers and turbines where it is used. For this job to be done as well as possible it is obviously essential that the processes involved should be fully understood so that the right principles can be applied in the design and operation of the plant. Theoretical studies and practical measurements have to be made towards understanding all aspects and stages of the process starting with the flow of heat from the interior of the fuel rod or pellet to the outside of the can, from the can to the coolant gas or liquid and from the coolant to the water tubes of the boilers and on to the turbines. If the coolant is a gas, as in Magnox reactors and AGRs, the dynamics of its movements will have to be ascertained and understood: these include pressure and volume changes, and patterns of turbulent or laminar flow in different parts of the circuit. In addition, problems of acoustic vibration can, and do, assume great importance: in some of the large AGRs they have led to post-design and even post-construction changes having to be made because the difficulties were not understood early enough. If the coolant is a liquid that boils, as in SGHWR, there are added problems, and they are large ones, of understanding two-phase (liquid and vapour) flow and the precise mechanisms of formation and condensation of steam.

Theoretical studies with computers, no matter how carefully the mathematical models are set up, are not enough. As with the nuclear aspects of design where research and materials-testing reactors play a major part, so here it is necessary to study the problems of heat transfer and fluid flow on a practical basis. For this purpose engineering test rigs have been constructed in which to measure flow rates, temperature changes, etc. in a wide range of practical conditions, using where necessary actual or simulated reactor or boiler components. These might, for example, comprise a full scale electrically heated dummy fuel element in a reactor pressure tube, with provision for close study of the boiling process under operational conditions, as in the high-pressure rigs at Winfrith. It is not enough to confine these studies to normal operations: it is important too that behaviour under fault conditions should also be understood. Rigs are therefore in operation, for example, to study in close detail coolant flow in a fast reactor fuel element channel that has become partially blocked. The use of lasers to study turbulent flow has been developed specially to help in this particular investigation.

Corrosion

Closely allied to flow problems are those of corrosion, and special rigs have been built to study the problem both inside reactors and outside. Here again test rigs built on an engineering scale are used to simulate actual components and conditions. Chemists and metallurgists work hand in hand with engineers, for their problems are very closely interwoven: vibration and stress influence corrosion rates and so does radiation, while the tendency for corrosion to arise, and its progress, will be strongly affected by any impurities that may be dissolved in the coolant. Further, when different structural materials are in contact with one another, the tendency to corrode may be greatly aggravated. Special difficulties may have to be overcome where different materials are on the inside and outside of a boiler tube, for example in the secondary heat exchanger of a fast reactor.

Plutonium Chemistry

Chemistry and the development of chemical processes for fuel manufacture and recovery have been touched on earlier, with the implication that much specialized research had gone into these developments. A great deal of this is applicable also in fields outside the nuclear industry, but the chemistry of plutonium, however interesting it may be academically, is of little direct practical relevance outside the nuclear fuel industry itself. Nevertheless its study has contributed to chemical technology.

The first plutonium-containing fuels for DFR, and later for PFR, were prepared by 'wet chemistry' processes involving solutions of plutonium nitrate from which the plutonium was precipitated by ammonia and heated strongly to form the oxide. In this series of processes the plutonium had to be handled successively as a soluble solid, a liquid and a heated dry powder. A new development which would reduce the dust hazard from such processes is the so-called 'gel precipitation' method. This was developed, first at Harwell and later in conjunction with BNFL at Windscale, and it avoids going through the heated dry powder stage, where the most stringent precautions are needed against inhalation. The mixed plutonium and uranium solution, containing a protective colloidal additive, is fed in controlled drops into a precipitant solution where each drop forms a sphere of hydrated oxide. These are dried and heated to give uniform spheres of mixed plutonium–uranium oxide of predetermined composition and very closely controllable size, density and porosity. These spheres, typically about half a millimetre in diameter, can be used directly by filling them into fuel element tubes: two different sizes of sphere used together permit extremely close packing in the tube and therefore very high density; or a quantity of spheres can be pressed together and sintered into pellets. This process of gel precipitation, like so many other processes developed for the nuclear industry, has numerous potential applications elsewhere.

Safety

As nuclear power has taken on increasing importance, a system of licensing and control has been evolved for the design and running of nuclear reactors, processing plant and laboratories, having particular reference to safety and to the disposal of radioactive wastes. This is administered and enforced by Government departments and agencies concerned with public health, safety at work and the protection of the environment. It has been applied with great attention and effectiveness, far more so than for any other industry, and is now regarded by many as setting a pattern which should be widely followed.

The guiding principles underlying the protection of people from radiation are based on the recommendations of the International Commission on Radiological Protection (ICRP). This is a small body of scientists nominated on the basis of their individual merit by the International Congress of Radiology, which is a professional body of long standing and unquestioned worldwide repute. The ICRP is thus answerable only to the world's professional radiologists and is totally independent of national or governmental interests and influence. Its findings and recommendations on radiological protection and the limits of radiation dose are accepted and used by all major countries of the world, including the United Kingdom. The guiding principles adopted in the UK are based on the ICRP recommendations. They put the onus on the management of a nuclear installation to do much more than merely keep radiation emissions below a set maximum—they have to show that they really are trying to cut emissions down as far as is reasonably possible, taking all the circumstances into consideration. Radiation dose limits are set at substantially higher levels for small population groups such as people working in the nuclear industry itself. This is because these people are subject to medical selection and periodical check-ups, their working conditions are closely controlled and their numbers are too small to be statistically significant as far as the genetic dose to the population is concerned.

To put the matter in perspective, the nuclear industry is responsible for well under one per cent of the total radiation exposure of the average member of the population of the United Kingdom: the great bulk of radiation exposure is from natural sources, rocks, cosmic rays, etc., with medical uses of X-rays and radioactivity the second largest. By far the greatest part of the small contribution by the nuclear industry is due to radioactive wastes disposed of to the environment: a small proportion of these find their way, mostly through plants and marine organisms, into the human food chain. The amounts involved are minute and the internal radiation dose that any of them gives is far below the relevant limits set by the ICRP. Most nuclear wastes are stored within the bounds of the industry's own sites, but authorized waste discharges are calculated by reference to ICRP limits for each radioactive species.

The safety record within the nuclear industry is exemplary, largely because the nature of its major hazard, radioactivity, is very well understood and has perforce to be treated with respect. The science of health physics has grown up as a vigorous hybrid between biology and the physics of radiation, concerning itself with the establishment and implementing of safe working practices. Under British law the operator of a nuclear plant is under an absolute obligation to see that nothing involving radioactivity that happens on his site, or to materials in transit or discharged as waste, causes personal injury or damage to property. He must carry insurance cover against this up to £5 million, and claims may be made against him up to 30 years after an alleged event. For these reasons if for no other he will be under a strong inducement to ensure that his plant is of the utmost safety and reliability, and he will employ his own specialist department or consultants to look after this aspect. One such specialist body is the Safety and Reliability Directorate of the UK Atomic Energy Authority (previously the Safeguards Division of the Authority's Health and Safety Branch) which functions independently of all the operating management units and acts in an advisory capacity to Government and the nuclear industry. Furthermore, before a start can even be made on building a commercial nuclear plant, the operator has to satisfy an independent Government watchdog body, the Nuclear Installations Inspectorate, on the safety aspect of every relevant detail of the plant's design and operating procedures.

The Future

The 25 years of our survey period has seen a continual, if uneven, increase in the amount of electricity produced by nuclear means, both in absolute production figures and in the proportion of nuclear generation in relation to the total.

The future scale of thermal reactor installation in the UK is not at present predictable, but unless the economy stagnates or unless the tide of public opinion swings strongly against nuclear power as such (rather than, as is happening now, only against certain aspects of its growth), then there is likely to be a substantial programme of thermal power station construction in the 1980s and 1990s. This will be based on whatever design of thermal reactor is considered appropriate at the time that design work has to start.

Britain is without question one of the world's leaders in fast reactor development. This enables her to benefit from international co-operation and technical exchange agreements, because she has a lot to offer in return. She is in a position to start designing and building a commercial sized fast reactor with associated fuel and reprocessing plant as soon as Government give the go-ahead to the industry. If fast reactors are to play their part in meeting the nation's energy demands at the turn of the century, when oil and gas will be getting scarcer and ever more expensive, then it is clearly necessary that the system should be proved at full scale in good time.†

THE QUEST FOR FUSION POWER

Fission and Fusion

In nuclear fission, the basic process of nuclear power generation as described in the preceding pages, the total mass of the particles resulting from the fission of a heavy nucleus is smaller than that of the parent nucleus. There is therefore a release of energy corresponding to this loss of mass. Similarly when light nuclei fuse together to form heavier nuclei there is also a loss of mass which appears as kinetic energy of the resulting nuclei. Unlike fission which rarely occurs in nature, fusion is one of the universe's major energy processes, happening on a vast scale all the time in the sun and the stars. However, on earth, fusion is more difficult to achieve than fission. The reason is simple: fission is caused by neutrons entering the nuclei of heavy atoms, which they can do without having to overcome any barrier of mutual repulsion due to electrical charges. They can, as it were, just drop in. Atomic nuclei, on the other hand, all carry positive electrical charges so they repel one another increasingly as they get closer together. Particle accelerators can readily produce streams of ions moving fast enough for a range of fusion reactions to take place in target material, but the energy used is much greater than that liberated because most of the energy of the impinging ions is dissipated in collisions with the electrons of the target atoms. However, work with accelerators has established which of many possible fusion reactions are likely to show the best energy yields: the most promising of these is that between the nuclei of deuterium and tritium (isotopes of hydrogen), which

†A public enquiry on fast reactors has been promised before construction of a full-size reactor is authorized.

react to give helium and a neutron and energy. Deuterium is plentiful because it is present in water, and tritium can be generated in a secondary nuclear reaction between the emitted neutron and lithium. Thus if fusion can be realised, it offers a vast new source of energy for man's use.

If the fusion process is to be useful, much less energy must be expended in getting the nuclei to fuse together than is liberated by the fusion reaction itself. What this means in practice is that the atoms must be stripped of their electrons and the matter then takes the form known as 'plasma', which consists of an electrically neutral mixture of positively charged atomic nuclei (ions) and electrons. It is the state in which most of the matter in the universe exists. If the plasma is at a sufficiently high temperature, of the order of 100 million degrees K which is several times hotter than the centre of the sun, the ions are moving about at random so rapidly that their thermal energy alone is sufficient to enable them to overcome their mutual repulsion, and 'thermonuclear' fusion can occur at a useful rate.

Plasma has some of the properties of a gas in that it diffuses and it can exert or respond to pressure, and some of the properties of a metal in that it is a good conductor of electricity and its movement can be controlled by magnetic fields.

Containment

Plasmas at a temperature of even a few tens of thousands of degrees K cannot be confined by a material vessel alone. No solid structure can exist at these high temperatures and in any case contact with the vessel walls would result in cooling of the plasma. In the sun and the stars the job of containing the hot plasma is done by gravity but this approach is of no value on earth since the plasma, being small and of low density, has a small mass. There are two broad lines of approach: firstly, 'inertial' containment of the plasma at a relatively high density for a very short time, and secondly, magnetic containment at a lower density for a longer time. Inertial containment—heating the plasma so quickly that it has no time to get away—is used in the H-bomb, where an 'imploding' fission bomb compresses and heats the reacting fusion materials strongly enough for a thermonuclear reaction to occur before the materials are dispersed. Such an approach, but using high powered laser beams for producing the implosion of a very small pellet of material, is being actively studied now in various countries with the aim of producing controlled fusion. However, from the early days of fusion research, much the most studied way of containing equally hot but less dense plasmas for the longer times has been by magnetic forces. This still forms the greater part of fusion research efforts, particularly in Britain, and is the subject of this chapter.

Magnetic Confinement

The electrically charged particles of which plasmas consist cannot cross the lines of force of a magnetic field nearly as easily as they can move along them while spiralling around them in corkscrew fashion. Therefore in a plasma held by a field of any given shape, there is much less tendency for the particles to escape across the direction of the field lines than along them, especially if the field is very strong. At the same time the faster the particles are moving (i.e. the hotter the plasma) the better able they are to force their way out across the field lines, and the stronger is the field needed to hold them in place. The shape of the confining field, and the rate at which its strength or direction changes, also greatly affects its ability to contain the hot plasma.

At the start, two magnetic configurations seemed (and still seem) to be possible: open-ended systems and closed toroidal systems in which the shape is like a motor-tyre tube or an American doughnut. In the open-ended system the ionized gas is contained by magnetic fields in a vacuum-tight straight tube. However, in such a system the field lines do not close on themselves within the vacuum chamber but intersect the end walls. The plasma following the field lines thus also goes to the end walls. In toroidal systems the magnetic field lines close on themselves within the torus and the escape of plasma is much reduced.

Heating

At the beginning of the research work the required temperatures of around 100 million degrees K, though at first sight appearing dauntingly high, did not look to be wholly unattainable, provided that the associated containment problems could be solved. There seemed to be no reason why very large electric current pulses passing through the conducting plasma should not be able to heat it to a temperature of several million degrees K. However, as plasma resistance decreases with rising temperature, at very high temperatures the method becomes less effective. Methods of heating not suffering from this disadvantage have since been developed; one used on DITE is described later.

Diagnostics

The techniques of plasma diagnostics have grown up from the start alongside those of heating and containment and have been an essential factor in enabling progress to be made towards fusion conditions. One of the most important plasma parameters to be measured is temperature, for which new techniques had to be developed. An early way was to study the visible spectrum of the light emitted by the hot plasma. By measuring the relative intensity of spectral lines an estimate of the electron temperature could be made. The extent to which spectral lines, normally of a single wavelength, were broadened by the 'Doppler effect' due to the rapid movements of the emitting ions (an effect better known in the realm of sound) gave a measure of the ion temperature. The ions often have a different temperature from that of the electrons. Recently new and more accurate methods have been developed using the light scattered by the plasma from a high-power laser beam to give both the electron temperature and density. Other measurements have included those of plasma pressure, magnetic field strength and direction, X-ray emission to give the electron temperature, neutral particle energy to give the ion temperature, neutron production, plasma position, plasma losses, and the build-up of impurities in the plasma from the vessel walls.

Problems

At the start of the 25-year period, it was clear that a great many years of intensive research would be needed to overcome formidable scientific and technological problems before a fusion reactor could be operated. Four main stages of development were foreseen:

1. Create and then contain a hydrogen plasma for a time sufficiently long for it to be heated to a temperature where nuclear reactions occur, as indicated by the emission of neutrons.

2. Raise the temperature further to between 50 and 100 million degrees K and hold it there long enough for the energy from the thermonuclear fusion reaction to more than balance the energy fed into the plasma. Theory indicates that this should be about one second for a plasma having a density of about 10^{14} deuterium and tritium ions per cubic centimetre—about equivalent to a good industrial vacuum. The time and particle-density requirements are related, in the sense that it is their product rather than either one individually that matters.

3. Construct a large apparatus to show that a net electrical power output can be produced reliably.

4. Develop the technology of the reactor sufficiently for the system to be able to compete economically with other power generating systems.

So far only the first stage has been studied. The coming into operation in a few years' time of large 'tokamaks' such as JET will enable a start to be made on the second stage.

Early Work

Since about 1947 work had been going on in the University departments in London (Imperial College), Liverpool and Oxford. By the early 1950s small straight-tube and toroidal devices had been built and the results published. Later, for security reasons, the work came to be concentrated at AERE Harwell, at the Associated Electrical Industries Limited Research Laboratories at Aldermaston Court and at the Atomic Weapons Research Establishment (AWRE) Aldermaston.

Work continued in secret for several years until 1956, when the leading Soviet nuclear energy scientist, Academician I.V. Kurchatov, was visiting Harwell and gave a lecture on the possibility of producing controlled nuclear fusion reactions in a gas discharge, describing the apparatus used and the state of progress and identifying some of the problems lying ahead. The work that he described was seen to complement British work. The wraps of secrecy were progressively removed and international exchanges were opened up between Britain, the USA and Russia.

It was clear that the outstanding problem confronting all fusion researchers was the stable containment of very hot plasma in a magnetic field: the thread of plasma, pinched by its own magnetic field, was found not to stay in place but to wriggle about violently, very soon colliding with the walls of the container. Experience with small toroidal devices had already indicated that size was of great importance, and that a large apparatus should be more successful than a small one in countering the problems of instability. Accordingly, ZETA (Zero Energy Thermonuclear Assembly) was built at Harwell and commissioned in the summer of 1957, specifically to study these instabilities, but also in the hope of achieving thermonuclear fusion reactions.

ZETA consisted of a toroidal tube one metre in bore and three metres in major diameter, with an iron transformer core, the primary windings of which were on the torus. The plasma acted as the single secondary winding of this transformer, carrying an induced current pulse every time the primary windings were energized. The magnetic 'pinch' effect of this current was controlled by entrapping within the plasma a toroidal magnetic field produced by additional coils around the torus (figure 6). This additional field led to improved stability of the plasma. Large banks of capacitors were used to feed controlled electric pulses into the primary windings which produced currents of up to 200 000 amperes in the plasma, later increased to a million amperes. As well as producing part of the confining magnetic field this current heated the plasma to temperatures of around two million degrees K. An appreciable number of neutrons were detected, suggesting that nuclear reactions were taking place. However, the problem of the wriggling plasma remained.

In January 1958 the work of ZETA was published, along with similar work being carried out in the United States, and ZETA was thrown open to inspection by journalists. The emission of neutrons had been taken to indicate that some success might have been achieved in bringing about thermonuclear reactions and this was taken up by the media and an alleged, but premature, promise of 'unlimited cheap power' was given a blaze of publicity. However, when it was discovered that neutrons could be produced by a small proportion of the ions being accelerated to high speed around the plasma ring and colliding with plasma ions, rather than from true thermonuclear (random) collisions, a deflationary and even hostile Press reaction set in. However, far from being made extinct by this setback ZETA was in fact at the threshold of a ten-year long and most honourable career, during which it formed the subject of over one hundred scientific and technical papers.

The Culham Laboratory

At about this time (1959–1960) decisions were being taken about concentrating all UK fusion research in one place where the security restrictions which surrounded much other atomic energy work need not be applied. This meant moving to an 'open' site which scientists and engineers from foreign countries would be free to visit. Culham, some eight miles from Harwell and Oxford, was eventually selected, and a range of purpose-built laboratories, workshops and ancillary buildings was designed to house the kinds of apparatus and services that future fusion research seemed likely to call for. In 1964 most of Britain's fusion research teams and their apparatus had become concentrated at Culham. ZETA, however, remained at Harwell where it was to continue in operation for a further four years until its final closure in September 1968. It was too large and complex, with all its ancillary plant and services, to make it worth moving. Nevertheless, it was still producing a steady output of useful scientific information, even though it had long been clear that it was never going to turn into a fusion reactor. It had demonstrated the stabilizing effect on plasma confinement of 'sheared' magnetic fields, that is to say, of magnetic fields in which the lines of force do not simply encircle the plasma as a garter encircles a leg, but lie at an angle to the axis of the plasma that changes with the distance from its axis, more like the strands in a rope. In this way the escape of plasma is discouraged: although at one radius it may be able to push 'between' the lines of force of the magnetic field, at a different radius it will have to cross them, which is more difficult. A period was found during the decay of the plasma current pulse where the outer part of the plasma was remarkably quiescent and where stability (freedom from wriggle) and containment (prevention of escape) were both improved. By the time of its closure in September 1968, ZETA was producing a confined plasma at a temperature of 2 million degrees K by means of a self-stabilizing discharge. It had therefore established that the toroidal configuration could, with further developments, become a serious contender for a fusion reactor design.

Culham's main initial task was to extend the understanding of the physics of hot plasmas contained by magnetic fields in both linear and toroidal configurations. On the theoretical side, special attention was paid to the use of computers for design calculations, for the interpretation of experimental results and for the construction of mathematical models of plasma confinement systems. A further field of importance was that of plasma diagnostics. A group concerned with the spectroscopy of high temperature plasma formed the nucleus in 1968 of the Astrophysics Research Division (based at Culham) of the Science Research Council's Appleton Laboratory.

A conscious effort was made to keep up with overseas work by exchanges of information, by visits, including long-term exchange visits, and by full participation in international Conferences and Summer Schools. All this helped to ensure that Culham established and held a position as a 'centre of excellence' in its own field while retaining effective contact with all the most advanced work elsewhere.

In 1962, scientists from the UK, the USA and the Soviet Union gathered at Harwell to exchange views and experience on plasma instabilities, and in 1965 the International Atomic Energy Agency's second conference on fusion research was held at Culham. In 1969 Culham was host to the first international conference on fusion reactor studies. In the same year, a Culham team, already well known for its work on plasma diagnostics, was invited to carry its experience into the Kurchatov Laboratory in Moscow. Some remarkable values of electron temperature had been claimed by the Russian researchers in their new 'tokamak' toroidal plasma experiments. The Culham team was able to confirm these results by an independent method of measurement.

In 1973 Culham's fusion research became part of a co-ordinated research programme under the auspices of Euratom. Under the Contract of Association, Euratom pays part (about 25%) of the cost of Culham's fusion programme and shares in its management.

Studies at Culham

Up to about 1965 the studies were concentrated on linear systems (apart from ZETA) but subsequently these gradually gave place to work on toroidal systems.

Open-ended (Linear) Systems

Because of their greater simplicity these are usually cheaper to build than toroidal systems. The simplest of Culham's open-ended systems was the 'Thetatron' for studying instabilities in plasma confined by the magnetic field produced by a single-turn coil 8 metres long wrapped round the length of the tube. The tube was long enough to make the end losses for the central region negligible during the lifetime of the plasma. Open-ended systems are also well suited for the study of the potential value for containment of a variety of magnetic field configurations and these were extensively explored at Culham. Thus, if the magnetic field in an open-ended system is increased at each end relative to that at the middle of the system a large proportion of the charged particles travelling towards the ends is reflected back towards the middle by the stronger field of the 'magnetic mirror' (figure 5). A further development of this

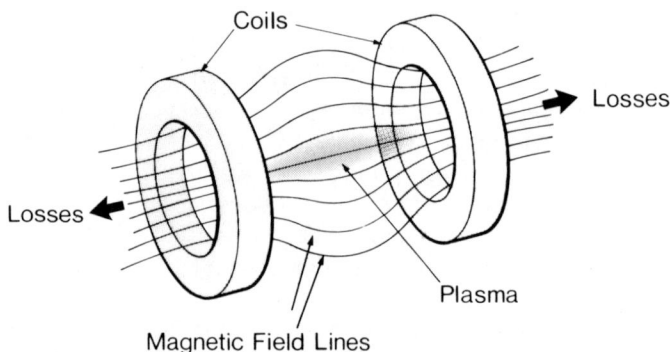

FIGURE 5.   Open-ended system: the magnetic mirror.

system is that of the magnetic 'trap' or 'well' in which the field strength increases in all directions outwards from the initial location of the plasma, so that the further the plasma gets from its origin the more strongly it is pushed back. The Phoenix II and MTSE I and II magnetic well experiments were used at Culham for these studies.

Toroidal Systems

Present experiments at Culham fall into four main classes, depending on the way in which the magnetic fields are produced. These are tokamaks, reversed field pinches, stellarators and the Levitron. In a toroidal system there are two main components to the field: the toroidal field in which the lines of magnetic force run parallel with the main circle of the plasma, and the 'poloidal' field in which they bind it around like a bandage. Figure 6 illustrates these fields and the basic components of the tokamak, reversed field pinch and ZETA systems. The relative strengths of the two fields determine the configuration of the resultant

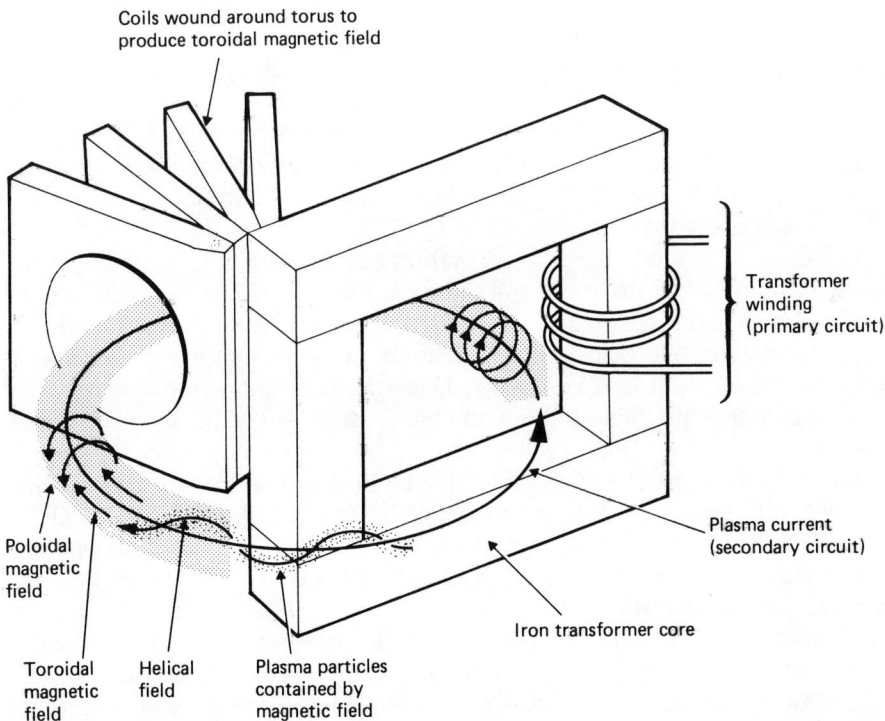

FIGURE 6.   Toroidal system: schematic diagram showing the main parts of the ZETA, tokamak and reversed field pinch systems. Only some of the toroidal magnetic field coils are shown; they are uniformly distributed all round the plasma ring. The transformer primary winding is sometimes wound on the toroidal vacuum vessel as in ZETA. In some machines the iron transformer core is omitted and an air core used.

field and its effect on the behaviour of the contained plasma. In the tokamak and reversed field pinch the configuration of the magnetic field is dependent upon the current flowing in the plasma which is not entirely stable in position; in the stellarator and Levitron the field is determined mainly by currents flowing in rigid coils and the magnetic configuration is therefore better defined.

(a) *Tokamaks.* In a tokamak the strong toroidal component of the field is produced by a set of coils wound around the toroidal vacuum chamber, while the weaker poloidal field is produced by an electric current generated in the plasma itself by transformer effect. This current also serves to heat the plasma. The tokamak system is a Russian development which has much in common with ZETA. For a number of years—ever since the 1969 visit by Culham scientists to Moscow—tokamaks have shown the greatest promise for further development towards fusion reactor conditions. Culham has two tokamaks, DITE and TOSCA, and CLEO which can be operated at will either as a tokamak or as a stellarator so enabling the two systems to be directly compared in one apparatus.

DITE (Divertor and Injection Tokamak Experiment) shown in figure 7 (plate IV) is unique in having provision to clean up the plasma during operation, by magnetically scraping off the outermost layer of plasma by a magnetic divertor. It also has an arrangement to supplement the resistive heating effect of the induced current in the plasma by the injection of powerful beams of fast-moving neutral atoms. These are produced by accelerating charged hydrogen atoms electrically and injecting them into the plasma after their charge has been neutralized by electrons. Here they become ionized again and collide with the plasma particles, substantially raising the mean kinetic energy, i.e. the temperature of the confined plasma. Plasma temperatures of 10 million degrees K have been routinely achieved.

TOSCA (Tokamak Shaping and Compression Assembly) is a small versatile tokamak designed to study ways of increasing the plasma pressure that can be confined by a magnetic field of given strength. The ratio of plasma pressure to magnetic pressure is called 'beta'. Beta is of great importance because it indicates the effectiveness with which the magnetic field is being used, and it is the magnets that are likely to represent the greatest part of the cost of a fusion reactor.

(b) *Reversed Field Pinch.* HBTX I (the High Beta Toroidal Experiment) has achieved high beta ratios by using magnetic fields similar to those used in ZETA but with a controlled reversal of the toroidal field direction between the plasma and its surrounding wall: this gives a very high shear to the magnetic field and leads to improved containment.

(c) *Stellarators.* Stellarators have fields with the toroidal component provided by current in coils spaced at intervals around the vacuum vessel, much as in tokamaks, while the second confining field is produced by several pairs of helical conductors wrapped in an open shallow-pitched spiral round the whole circumference of the vacuum vessel. The resultant complex magnetic field takes the form of nested closed circular tunnels of roughly triangular section with the plasma lying inside and separated from the wall. Stellarators offer a possibility

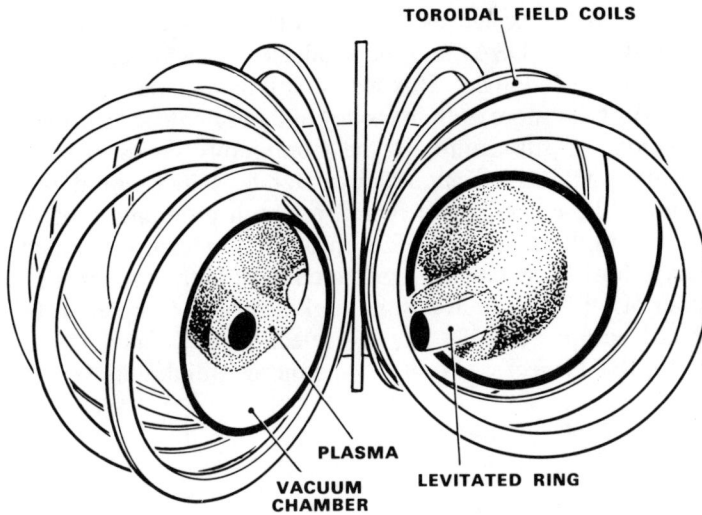

FIGURE 9. Schematic diagram of the Levitron.

of a continuously operating (as distinct from pulsed) fusion reactor. Figure 8 (plate V) is a photograph of plasma contained in the TORSO stellarator in which the containing magnetic field is produced entirely by currents flowing in the helical conductors visible in the figure.

(d) *The Levitron.* In this, the poloidal field is produced by a current that passes through a solid ring-shaped conductor immersed in the plasma, rather than through the plasma itself (figure 9). This current is akin to that induced by transformer action in tokamak plasmas, but in this case, as the current cannot produce plasma heating, neutral injection or high-frequency electric field heating is employed. In the superconducting Levitron at Culham the plasma is contained between the inward-pushing field produced by external coils around the toroidal vacuum chamber and the outward-pushing field produced by the current in the central core. It therefore takes the form of a toroidal tube, analogous in some ways to the inner tube of a car tyre which is held in place by the opposing pressures of the cover and the contained air. The conductor is a cooled ring of titanium-niobium superconducting alloy 0.6 metre in diameter which is held in position without mechanical support by the magnetic field of the current flowing in the conductor. Once induced this current flows for 15 minutes, allowing ample time for experimental plasma studies. The Levitron is not intended to be an approach to a fusion reactor but rather a basic experiment in plasma confinement in an ideal toroidal configuration.

JET (Joint European Torus)

In 1971, discussions began within the EEC on a proposal to build an experimental fusion apparatus, based on the increasingly popular tokamak concept, which would bridge the gap between the largest apparatus at that time being planned (in the USA) and the smallest size in which true fusion reactor conditions might be attained. This would still have to be a very large machine, probably costing far more both in money and in scientific effort than any single member-nation, or sub-group, in the Community could afford. By 1973 a decision had been taken to design such a machine, which would form the biggest single item in the co-ordinated fusion research programme of the European Communities. In the same year a European design team assembled at Culham, which played host to the team and provided most of the supporting services. By the summer of 1975 the main design work had been completed, but it was not until the late autumn of 1977 that the selection of Culham as the site for the international JET Project was finally announced. Building is expected to start in 1978.

Towards a Fusion Reactor

Parallel with all the abstruse but promising development work in plasma physics and magnetic confinement systems, in computer modelling, in diagnostics and in all the other necessary lines of research, increasing attention is being paid to the down-to-earth problems of designing, building and operating a fusion power station. A Culham team has produced its own notional design for a 2000 MW(e) reactor based on studies of what is practicable in the light of existing technology and its reasonably foreseeable developments. There is as yet no certainty as to which of the major systems of magnetic field configurations will ultimately turn out to be the best. However, some form of tokamak is the present favourite, and this has been accepted as a basis for design by the Culham team. Knowing that this could be subject to revision they have concentrated on matters common to a variety of configurations and systems. These include constructional materials, vacuum systems, modular construction of the toroidal vessel, superconducting magnets, and the molten lithium blanket system which will act as the combined neutron absorber, tritium breeder and heat transfer medium. Other important items include repairs and maintenance of the vacuum vessel and blanket, radiological protection, reactor accident characteristics, waste problems, fuel procurement, etc.

The fuel for fusion is plentiful: deuterium is present in the waters of the sea in quantities which will ensure its abundant availability for as long as Man is likely to survive. Tritium will be bred in the reactor blanket by the reaction of lithium with neutrons. Lithium is by no means a rare element, and in the long run it may not be essential because fusion reactors may be designed to run at a higher temperature on deuterium alone. But whatever the fuel, it is abundantly clear that a fusion reactor is likely to be a very large and complex structure and correspondingly expensive to build even if assembly line production becomes possible. Power may become plentiful, but fusion power stations are never likely to be cheap.

RADIOISOTOPES—FRUITFUL BY-WAYS IN NUCLEAR RESEARCH

If progress to cheap and plentiful electric power represented the main road of advance for nuclear energy during these Jubilee years, then the production of radioactive materials—radioisotopes—and their applications to practical ends represented a maze of by-ways leading to some very fruitful fields. British science and technology—and business acumen—have played an active and successful role in these developments.

Long before nuclear fission was discovered the nature and properties of radioactivity were well known, and something of its potential value was evident, even if, because of nature's parsimony over supply, it was only realisable in a very limited field. With radium as the only available radioactive material, but at a cost of over £8 a milligramme, vital medical treatment was almost the only field in which it was used, the other being in self-luminous paints and markers. The scientific value of radioisotopes for studying chemical reactions had, however, been known for many years. Before World War II started, small amounts of a few radioisotopes were being made with the aid of charged-particle beams from accelerators, but it was the discovery of nuclear fission in 1938 that opened the possibility of supplying man-made radioisotopes in quantity and variety. However, it was not until after the war was over and the necessary combination of technical effort and reactor facilities became available, that hopes began to be realised.

In Britain, one of the first tasks of Harwell's two early reactors, GLEEP and BEPO, was to start producing short-lived radioisotopes for some of the already-known medical diagnostic applications. In October 1948 the Isotope Division was formed, and an isotope catalogue was issued listing radioisotopes of some 60 elements which could be made available by the irradiation (bombardment with neutrons) of suitable target materials in BEPO. This hitherto undreamed-of range of radioactive elements, available at moderate cost, marked the real beginning of radioisotope applications over a wide field of medical science. In particular it permitted the development of diagnostic techniques based on the use of radioactive tracers, where a harmlessly small, but easily detectable, quantity of a radioisotope of a particular element is used to 'label' and follow a much larger amount of the same element in its movements within the body. An early example was the monitoring of blood-flow in skin grafts by intravenous injection of radioactive sodium 24 as normal saline (salt solution similar to that in blood), and its detection or not in the area of the graft by the gamma-rays that it emitted. No gamma-rays meant no blood-flow, giving early warning of a rejected graft.

Not all radioisotopes produced in BEPO (which soon superseded GLEEP in this role) were in a form directly suitable for the user: many had to be chemically purified, or converted to some other chemical compound, or isolated from the bulk of the target material in which they had been created. This was done not at Harwell but at the Radiochemical Centre at Amersham in Buckinghamshire, where a radium refinery had been in operation since 1940, managed by the private firm Thorium Ltd. In 1946 this had been reconstituted as the Radiochemical Centre, and in 1950 it became a part of the Atomic Energy Authority. In 1960 there was a substantial reorganization in which the entire radioisotope production and sales functions were made the responsibility of the Radiochemical Centre at Amersham, part of the Harwell team becoming the Isotope Production Unit of the Radiochemical Centre. Technical liaison and commercial relations with hospitals, and with research organizations, universities, etc. which were also beginning to make wider use of radiochemical techniques, became the responsibility of the Centre alone. Under the uninterrupted leadership of Dr. Patrick Groves, who was in charge at Amersham from 1940, it has continued to develop and expand during the whole period under review. The Centre has always enjoyed the co-operation and involvement of scientists and medical users and stands in the front rank of world suppliers of radioactive materials of the most diverse kinds and uses. Since 1971, The Radiochemical Centre has operated as a limited company whose shares are held by the United Kingdom Atomic Energy Authority on behalf of the Government. It has major subsidiaries in the USA and Germany, and an additional factory is under construction in South Wales to supplement the Amersham facilities. The most rapidly growing line is in 'kits' which permit simple but accurate measurements of extraordinarily small amounts of hormones and other biologically active substances in small blood samples. Such assay kits are now very widely used in routine diagnostic tests.

Clinical tests, important though they are, represent only a part of the medical radioisotope story. Radium and X-rays had long been used in the treatment of cancer, and it was evident that the new availability of a range of artificial radioisotopes would increase the scope and flexibility, and perhaps reduce the cost, of this form of therapy. And so it turned out. Cobalt 60 was pioneered by Harwell as a source of radiation for deep-ray therapy, and formidable design and construction problems of radiation shielding and dosage control were successfully tackled. Radioisotopes of gold, tantalum, iridium, iodine, phosphorus, etc. were produced to meet various requirements for internal or external radiation treatments, and here again the very closest co-operation was maintained with the medical world. Today cobalt-60 teletherapy (deep-ray treatment) units of British make, many of them charged with cobalt produced by the Isotope Production Unit in DIDO or PLUTO, are in regular use in the UK and overseas, and The Radiochemical Centre Ltd markets a range of other clinical radiation sources.

The introduction of radioisotope techniques into industry followed a very different course. Although the scientists of Harwell's Isotope Division (which retained responsibility for developing non-medical uses) were themselves fully convinced of the potential value to British industry of a whole range of radioisotope applications, they soon found that very few of industry's leaders were prepared to back them by actually adopting the techniques recommended. However, by developing close relations with those firms that were prepared to be interested, it was found possible to ascertain the real needs of industry in some of those fields where radioisotopes might help, and to develop techniques or hardware that could be clearly demonstrated as both relevant and profitable to the user. To do this, visits by scientists from the Isotope Division were made to firms, and reciprocal visits to Harwell were organized, conferences were arranged at local centres of industry, and the help of Government and other organizations such as the TUC, the Federation of British Industries and the Institute of Directors, was enlisted. Perhaps most effective of all, however, were the courses organized at the Isotope School. This had been set up at Harwell in 1951 partly to further the industrial drive but mainly to help familiarize the medical world with the actual use of radioisotopes in hospitals and laboratories. Besides doctors and industrialists, teachers, journalists, trade union officials, even members of Parliament, came to Harwell on courses ranging from half a day to several weeks.

One of the first and perhaps most immediately successful industrial techniques was gamma-radiography. Here a small pellet of radioactive material, suitably shielded in a heavy metal container, acts as a source of penetrating radiation for the examination of castings, welds or engineering components. It uses a photographic film in just the same way as an X-ray set-up in a hospital or factory. The radioisotope source, usually cobalt 60, iridium 192 or thulium 170, in its shield is fully transportable and is independent of power supplies. It is ready for immediate use and, at least in its latest developments, is in no way inferior to an X-ray machine in the quality of picture; also it can be used in confined or otherwise awkward places, e.g. under the sea or in an inflammable atmosphere. Gamma radiography was largely pioneered at Harwell and is now in successful use throughout the world's industries. It has been employed on such diverse subjects as Stonehenge, the engines of jet aircraft, and undersea oil pipes.

Another early and successful radioisotope development was in the use of the rather less penetrating beta-radiation (fast-moving electrons emitted in radioactive decay) in non-contacting gauges for measuring, and hence for controlling, the thickness, density or weight per unit area ('substance') of continuously produced materials in sheet, strip or similar form. Gauges were developed through the joint efforts of the user industries (paper, plastics, rubber, sheet metal, etc.), instrument firms and Harwell, and they were brought to such a state of effectiveness that they are no longer regarded as novel or unusual. It is interesting to note that in one application, the continuous control of packing density of tobacco in cigarettes, the whole of the initiative and technique came from a firm that specialized in cigarette-making machinery.

The absorption, scattering and degradation of 'radiations' within matter, and the stimulation of secondary (fluorescent) radiations, follow very complex rules and it has been possible to take advantage of these in developing a range of outwardly simple analytical instruments. In these, a radioisotope 'source' produces radiation which interacts with the specimen under study in such a way that the radiations reaching an associated detector carry with them information about the composition of the specimen. Over the years, and as a result of close co-operation between interested parties, a number of instruments have come on to the market, some of them tailor-made to do a particular job such as measuring the thickness of tin plate on steel or the non-combustible ('ash') content of coal in its passage through a washing plant, and others for more general purposes such as the *in situ* assay of any one of a range of metals in an ore body or collection of prospector's specimens.

Development of this type of instrument to the stage of commercial acceptance involves long and close co-operation between representatives of many different interests: for example, the development of an instrument for 'logging' boreholes drilled while prospecting for copper used the pooled skills and experience of geologists, prospectors, mining engineers and instrument manufacturers as well as of Harwell scientific staff specializing in radioisotope instruments. The differing, and sometimes conflicting, interests of each of these parties has to be reconciled in the commercial as well as the technical aspects: the instrument must not only do its job, and continue to do it under conditions that to the laboratory scientist may seem quite unreasonably severe, but are commonplace to the user, it must also sell at a realistic price and (for such is business) bring in commercial profits all along the line. Indeed, one of the major changes characteristic of the period was the increasing emphasis that came to be put on the commercial worth-whileness of going on with promising scientific ideas.

So far we have looked at ways of using radioisotopes that affect us as individuals either indirectly through improvements in the products of industries using them, or only seldom, when we are undergoing specialist clinical examination or hospital treatment. One application, however, is likely to be noticed favourably by all of us from time to time: that is the use of gamma-radiation for the bulk sterilization of prepacked medical equipment, in particular hypodermic syringes. This is based on the well-known capability of gamma-rays to kill organisms of all kinds without the application of heat or chemicals. In the late 1950s, plans were in hand in Harwell's Isotope Division to build a plant for sterilizing bulk quantities of pre-packed plastic hypodermic syringes, etc. by exposure to gamma-rays at about 500 times the lethal dose to man. The plant, known as PIP (Package Irradiation Plant), was built at Harwell's nearby Wantage Research Laboratory and, with the co-operation of medical authorities and firms making medical and pharmaceutical products, the process was developed and successfully demonstrated on a commercial scale. The syringes, etc. do not have to withstand heat so they can be moulded from inexpensive plastics, they are individually packed to remain sterile until required, and they are used once only and then discarded. There is no danger of cross-infection and every patient gets the benefit of a factory-sharp needle. Everybody is pleased, especially perhaps the doctor on his rounds, and the method is now in commercial use throughout the world. There is, of course, no question of gamma-irradiated materials becoming radioactive: in this respect gamma-rays are like light, and even the most over-ardent sunbather does not glow in the dark after a day on the beach—even if the damage he has done to his skin makes him feel as if he does.

The energy of radioactive decay can be used even more directly: a very significant use, again in the area of medicine, is in nuclear-powered heart pacemakers for patients suffering from 'heart-block'. Here a tiny pellet of plutonium 238, a bank of thermocouples and some very advanced micro-electronic circuitry are sealed in a plastic capsule about the size of a duck's egg and stitched into the body cavity. An electric lead carries pulses of current to the heart muscle, providing the regular stimulus that the patient's own nervous system can no longer supply. It will continue to work without recharge or replacement for many years: at the time of writing, seven years has been achieved in one patient and many more years of uninterrupted and effective use can be expected.

Other energy-producing applications of radioisotopes include light-sources ranging from telephone-dial illuminators to coastal navigation lights.

The tracer applications of radioisotopes, which have from the first been so useful in medicine, have proved as attractive in the fields of industry and public works and in environmental and agricultural research. For example, measurements have been made of the sea-bed movements of dredging spoils, the longshore drift of pebbles, the spread of industrial wastes in estuaries, the flow-rates of rivers in flood or drought conditions, the movements of materials through industrial processes or the life-cycles of plants or insects. Teams have gone out to many parts of the world to apply radioisotope techniques to local problems of water supply or of pollution, while nearer home the North Sea oil industry has used tracer and instrumental applications of radioisotopes in controlling the grouting of drilling platforms to the sea-bed, checking the integrity of structural welds below water-level, and locating leaks in undersea pipelines.

All applications of radioactive substances are subject to stringent regulations to ensure the safety of workers and the public. In the early days of their development there was a tendency among industry managers to worry overmuch about safety aspects, and later about the regulations themselves. But in the event, whatever the fears, radioisotope techniques have turned out, like nuclear power itself, to be exceptionally safe tools to work with and are now fully established in many fields.

THE JUBILANT ELECTRON

S.J. ROBINSON, F.R.S. (Editor)
M.E.L., Crawley, W. Sussex

INTRODUCTION

As electrical power engineering is concerned with the control of energy, so is electronics with the processing of information, and while there is a trend to large size to increase the capacity and efficiency of energy conversion machines, there is a corresponding trend to small size to increase the 'bandwidth' and information handling capacity of electronic circuits. The topical phrase 'small is beautiful' may therefore be applied to electronics in a real sense, and it is above all the reduction in the size of information processing circuits which has characterized progress over the last 25 years.

However, small size is not enough. Shannon first recognized that energy dissipation is the 'friction' of information processing and this makes it vital to design low power systems, if the information handling potential of small circuits is to be realised. Here the electron phenomena which occur in semiconductor materials at low voltages and currents have proved ideal for signalling the transfer of information.

The phenomena of semiconductor physics have been harnessed by photographic techniques with bewildering effect. Since the Second World War the volume of electronic circuits has been reduced by many million-fold. Moreover, it is a happy corollary that the reduction in the use of material which accompanies small size leads to low cost where the circuit is used in sufficient numbers to defray the design expenses. Thus high capacity information handling circuits have the threefold merit of small size, low power consumption and low cost.

Such has been the progress in circuit design that problems increasingly relate to the transducer which takes the information in and out of the electronic system, and the means by which information is transmitted from one distant point to another. These functions often limit the information handling capacity of the system and, to obtain greater 'bandwidth', more and more solutions to transducer and transmission problems are being found in the optical, rather than the radio, part of the electromagnetic spectrum. It is therefore no coincidence that, in the electronics exhibits selected for the Royal Society Jubilee Soiree, optical and microwave disciplines are seen alongside those of low temperature

113

and solid-state physics. Indeed it may be that in the next 25 years the marriage of optical and electronics science will produce applications as startling as those produced by solid-state and photochemical science in the last.

Six applications of electronics are discussed briefly here:

1. Uses of the Josephson effect, an electron tunnelling phenomenon observed at the junction of two superconductors, prepared by A. Hartland (National Physical Laboratory)

2. Photon to microwave—new components in a novel radar, prepared by K.W. Gray and P.W. Braddock (Royal Signals and Radar Establishment)

3. The application of electronic processing to microwave phase measurement for the precise navigation of aircraft, prepared by R.N. Alcock and A.R. Cusdin (Philips Research Laboratories) and P.K. Blair (Standard Telecommunication Laboratories)

4. The application of temporal variations of light (photon correlation), with, as an example, the application to the measurement of blood flow velocities in the retina, prepared by E.R. Pike (Royal Signals and Radar Establishment)

5. Wide bandwidth information transmission along glass fibres using light as the carrier, prepared by A. Hartley-Smith (Standard Telecommunication Laboratories)

6. CEEFAX, a method for the wide dissemination of data by television, prepared by A.M. Daniell (BBC).

The Josephson Effects and Their Applications

In 1962 B.D. Josephson, a post-graduate student at Cambridge University, predicted a number of effects associated with the tunnelling of pairs of electrons between two superconductors separated by a thin insulating barrier. These were soon confirmed experimentally and devices employing Josephson junctions of various types have since contributed to the advancement of several techniques of great importance in the field of electrical measurements. Three such applications currently under investigation at the National Physical Laboratory are briefly discussed.

The Josephson Effects

Josephson junctions are formed when two separate pieces of superconducting material are brought into very close proximity, either by inserting a thin insulating barrier between them, or by making one of the superconductors very narrow at the point of contact. For example, a lead–lead-oxide–lead junction is made by successively evaporating two overlapping lead films each with a thickness of about 150 nanometres on to a glass substrate. Before evaporating the upper film, the lower film is allowed to oxidize to a thickness of about 2 nanometres to form the insulating barrier. When such a junction (placed in liquid helium to ensure that the lead films are superconducting) is current biased, the dc Josephson effect is revealed by the current/voltage characteristics, shown in figure 1, plate I. A supercurrent flows up to a critical value I_c, a voltage only appearing across the junction when I_c is exceeded. The characteristic shown in figure 1 has considerable hysteresis typical of tunnel junctions having a fairly large capacitance.

In the non-zero voltage state besides an ordinary dc current Josephson also predicted that an ac supercurrent would flow in the junction whose frequency f_J is determined by

$$f_J = K \times V_J$$

where K, the Josephson constant ($\equiv 2 \times$ charge on the electron/Planck's constant), has the value 483 594.0 GHz/volt, and V_J is the voltage developed across the junction. The ac Josephson effect is revealed by irradiating the junction at a frequency f. Since harmonics of f become equal to f_J the current/voltage characteristic takes the form of a staircase of current steps the voltage separation of each step being precisely f/K.

Voltage Standard Monitoring System

Until recently all national standards of voltage were based on the mean emf derived from groups of saturated Weston cells. However, this emf (about 1.018 volts) is not constant with time. Consequently, to ensure that the maintained units of voltage for all countries did not drift apart it was necessary to hold a triennial intercomparison under the auspices of BIPM (Bureau International des Poids et Mesures). Cells, in temperature controlled enclosures, whose emf's had been calibrated in terms of the national standard were taken to BIPM at Sèvres, on the outskirts of Paris. Since the BIPM maintained volt was also based on a group of saturated Weston cells, this periodic intercomparison of national standards could not eliminate any drift relative to the 'absolute volt'. The 'absolute volt', i.e. a volt determined in terms of the basic units of mass, length, time and current has an uncertainty of about 4 microvolts, whereas the present demand arising from the stability given by standard cells is for voltage measurements having a precision of 0.5 microvolt or better.

Following the predictions of Josephson it is now possible to construct a voltage standard monitoring system whose precision depends, in principle, only on a frequency measurement. Such a system is shown in figure 2. At the heart of the system is a lead–lead-oxide–lead tunnel junction which is located in a section of wave guide and irradiated with a frequency of about 10 GHz. The junction is current biased on a step in its current/voltage characteristics such that the voltage across it is about 2.5 millivolts. A calibrated potential divider enables a direct comparison between the junction voltage and the emf of a standard cell to be made.

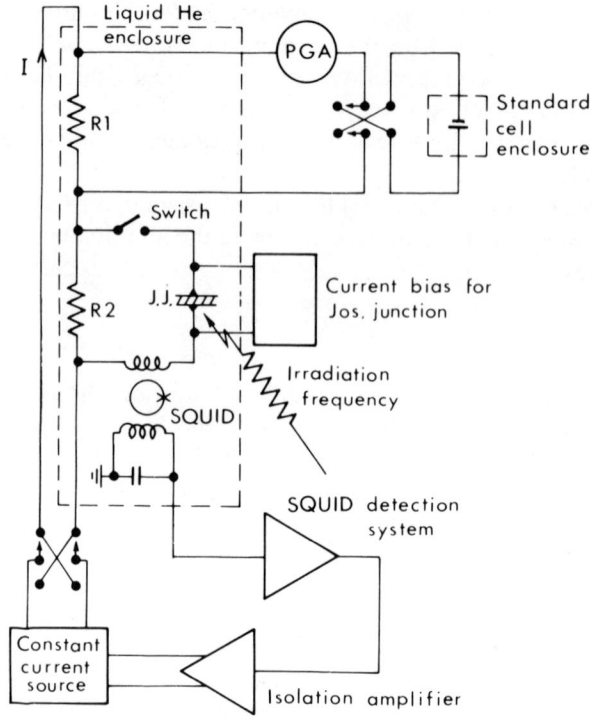

FIGURE 2. Block circuit diagram of Josephson junction voltage standard monitoring system.

The voltage V at the output terminals of the system is calculated from

$$V = \frac{p \times n \times f}{K}$$

where p $(= R_1/R_2)$ is a potentiometric factor which can be measured, K is the Josephson constant, n is a known 'step' integer and f is the frequency of the radiation applied to the junction. Since f can be measured with great precision (1 part in 10^{12} at the NPL) the main contribution to error lies in the determination of p. However, V (about 1.018 volts) can be determined with an uncertainty of 0.000 000 02 volt (20 nanovolts).

This system does not enable one to improve the accuracy for the determination of the 'absolute volt', and to achieve this K would have to be measured in terms of mass, length, time and current in a separate experiment. However, the relative drift of the mean voltage of the group of cells comprising the national standard can be monitored. Currently, at the NPL, this is -0.02 microvolt/month.

Josephson Junctions at Sub-millimetre Wavelengths

Another aspect of the interaction of low levels of incident electromagnetic radiation with Josephson junctions is a very nonlinear behaviour enabling them to be used for detection, harmonic generation and frequency mixing of radiation sources. These effects can be employed at frequencies up to at least several thousand gigahertz and there is particular interest in frequencies above 100 GHz (3 mm wavelength), where more conventional microwave devices often suffer a severe deterioration of performance.

The Josephson junction which is most sensitive at microwave and higher frequencies takes the form of a finely pointed superconducting wire, made from niobium, pressed against the flat surface of a niobium block. This type of junction has been used in heterodyne mixing experiments carried out with a 891 GHz (0.337 mm) HCN laser as the test signal. With a second, similar laser as the local oscillator the heterodyne (pre-detection) noise equivalent power is 10^{-17}W Hz^{-1} which is superior to that of conventional mixers with high intermediate-frequency bandwidths at this wavelength.

Using stable pre-set devices, an 18 GHz klystron can be phase-locked, using 50th harmonic mixing, to a 891 GHz laser. The laser frequency can then be determined by measurement of the klystron frequency. This technique was used as part of the recent NPL determination of the speed of light. The unprecedented harmonic-generating capability of a single Josephson junction is revealed by the multiplication up to 825 times of signals near 1 GHz and their mixing with the 891 GHz frequency. This is achieved with a relatively low input power (about 100 microwatts) at 1 GHz, derived directly by multiplication from a 120 MHz quartz crystal oscillator. Such experiments have provided accuracies of 1 part in 10^{10} or better for the measurement of frequencies in the far infrared spectral region.

The heterodyne mixing properties are likely to be of particular use in plasma diagnostic experiments on fusion type plasmas where sensitive wide-band receivers are required. Also, under consideration, is their application in submillimetre heterodyne astronomy where the requirement is for a receiver operating in the wavelength range 0.5 to 1 mm.

Experimental Quantum Standard of RF Attenuation

The RF SQUID (Superconducting Quantum Interference Device) is essentially a magnetometer whose flux state is sensed at radio frequencies. The active part of the magnetometer comprises a superconducting loop which contains a Josephson junction. In the microwave SQUID used in this application, the magnetic flux to be sensed by the SQUID is produced by passing current through an rf cable (figure 3). The phenomenon of flux quantization results in the SQUID experiencing a series of flux states as the current is increased or decreased.

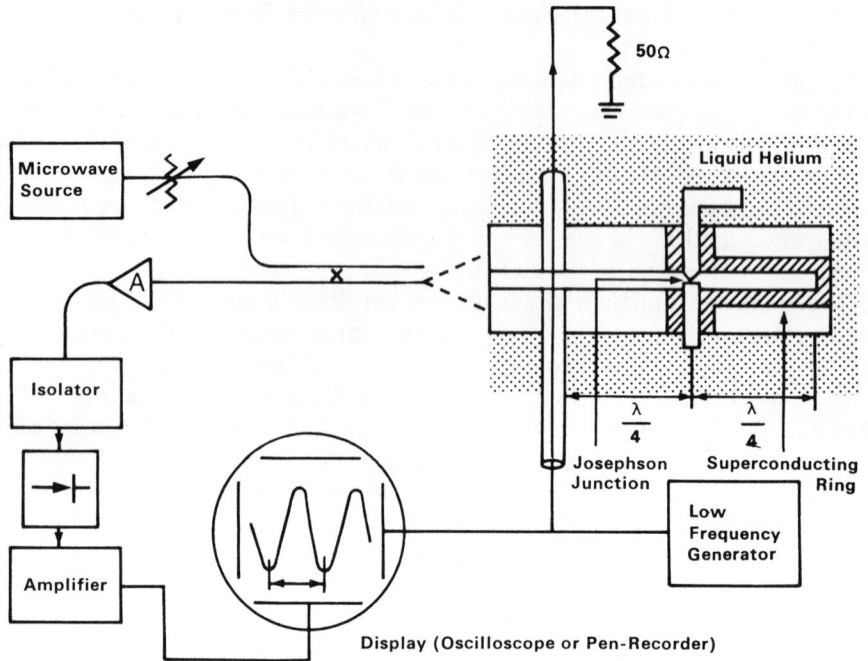

FIGURE 3.   Microwave SQUID construction details: circuit for monitoring low-frequency periodicity of the SQUID.

A linear variation of direct current produces a periodic variation in the 'interrogating' microwave signal reflected from the SQUID. The incident microwave power, the point-contact Josephson junction and the detecting system can be adjusted so that this periodic response is a near-sinusoid. Under these conditions the time-averaged response to increasing rf current can be shown to be a zero-order Bessel function. The zero crossings of this response are well defined and can be used as a scale of rf current ratios, and from this, in a constant impedance circuit, an attenuation scale can be defined. The SQUID attenuator is particularly attractive because it has a broad instantaneous frequency range; for example from dc to 2 GHz for an X-band SQUID. It is also free of the very precise machining requirements of other standard attenuators. Comparisons with a conventional waveguide beyond cut-off attenuator indicate a calibration precision of 0.003 dB.

Conclusion

The applications of Josephson devices have been extensive in many areas of physics and electronics and it is to be expected that this growth will continue into the future, particularly in the field of computation (for example, tunnelling cryotrons are being investigated as computer elements). It is hardly surprising that B.D. Josephson was awarded the Nobel prize for physics in 1973.

PHOTON TO MICROWAVES—NEW COMPONENTS IN A NOVEL RADAR

A solid state motion detector radar set, giving visual warning when any person approaches, is shown in figure 4, plate II.* It incorporates a number of new concepts, highlighting advances in British solid state technology. The use of a high efficiency solid state microwave transmitter and novel signal processing electronics has reduced the power requirements to such a degree that sufficient power can be obtained from the spiral array of solar cells, operating in normal room lighting. The visual indicator panels contain a new stable liquid crystal material. In this section a short description is given of the details of the overall motion detector and of the background of two novel components used, indium phosphide transferred electron oscillators and biphenyl liquid crystal displays.

Motion detectors and intruder alarms of the solid state microwave type are generally cw homodyne radar systems using the Doppler-shifted return signals that moving objects provide. They have found wide application in banks and traffic light control. Under operating conditions they consume several watts of power which is provided by a mains generated dc supply or in an emergency by a battery. There are a large number of applications where low power drain and therefore continuous battery operation are mandatory. To this aim pulsed-Doppler radar operation can be applied advantageously. The overall mean power consumption is then in principle simply controlled by changing the duty cycle operation in such a way as to ensure that sufficient return pulses are obtained within the highest Doppler frequency of the target. With the use of a gated receiver system the overall noise performance is optimized, thus enabling the exhibit to operate with no sacrifice in signal detection sensitivity compared with the cw type, even though the overall dc power consumption is 1000 times less. Details of the layout of the exhibit are illustrated in figure 5. The minimum transmitted pulse width is determined by the signal-to-noise ratio of the radar and depends on the aerial gain, pulse output power and noise figure of the receiver. The received signal after the mixer is integrated and stored between pulses, amplified, filtered, and then used to drive the liquid crystal display at the Doppler frequency. Since this radar system is only sensitive to signals that return from targets within the pulse length it is automatically self-range-gated. The major power consuming component is the transmitter and it therefore necessitates the use of an efficient component as exemplified by the indium phosphide transferred electron source. Complementary metal-on-silicon (MOS) circuits in the low frequency electronics and liquid crystal cells for the displays enable an extremely low power consumption to be achieved. Details on the overall radar performance are given in figure 4.

*The radar set is similar to that presented to His Royal Highness The Duke of Edinburgh on his visit to RSRE, Malvern, on 26 March 1976 (Plate III).

InP
T.E.O.

f_0

Tx

f_0

$\Delta f = \frac{2vf_0}{c}$

PULSED
MODULATOR

T

MIXER

Rx $f_0 + \Delta f$

TARGET VELOCITY
(V)

Δf

GATE

SIGNAL
PROCESSING

T

LIQUID CRYSTAL
DISPLAY

BLOCK DIAGRAM

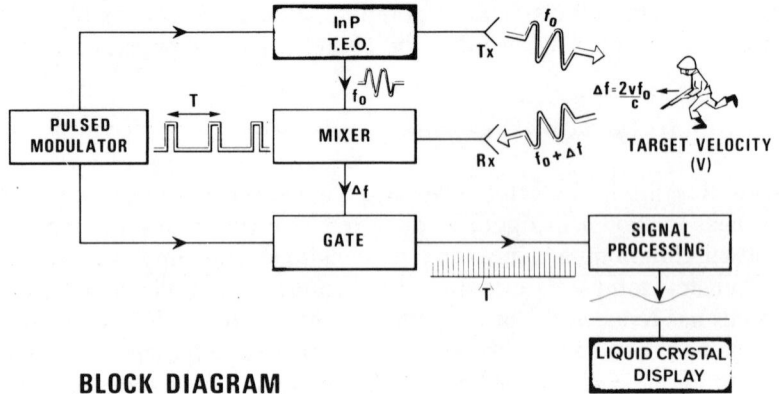

FIGURE 5.   Schematic of the radar electronics. T = interpulse period; f = frequency; Tx = transmitter; Rx = receiver; Δf = Doppler frequency; T.E.O. = transferred electron oscillator.

Transferred Electron Microwave Oscillators

Transferred electron sources made from gallium arsenide which have been under development in the United Kingdom over the last fifteen years are now widely used in radar systems, and the high efficiency indium phosphide equivalent will further extend this application. The conventional microwave vacuum devices, magnetrons and klystrons, of course will still have a role to play but are now concentrated almost exclusively in the high power field where many tens of watts and upwards are required.

The dynamics of the transferred electron effect, a phenomenon exhibited by certain compound semiconductors involves the transportation of hot electrons from a high mobility state to conduction states in which their mobility is low. The resulting negative differential mobility contributes to the overall 'active' behaviour which is exploited in the microwave operation of the material. In the classical experiment by Gunn, in 1963, coherent high-frequency oscillations were first observed with gallium arsenide when biased above a threshold voltage level. It was soon realised that since the frequency of operation was inversely proportional to the semiconductor thickness, epitaxial growth techniques could be used for producing the very thin semiconductor active layers of a few micrometres suitable for operation up to frequencies of several tens of gigahertz. With the immense possibilities the Gunn device seemed to offer, the simplicity of its two-terminal structure and the readily available techniques for materials growth, many firms became involved in its manufacture. The success of GaAs as a microwave material is very apparent today but it was recognized at a fairly early stage that it was not the optimum choice of semiconductor for exploiting the transferred electron effect and that improved performance in terms of both microwave power and efficiency would be obtained from a number of other

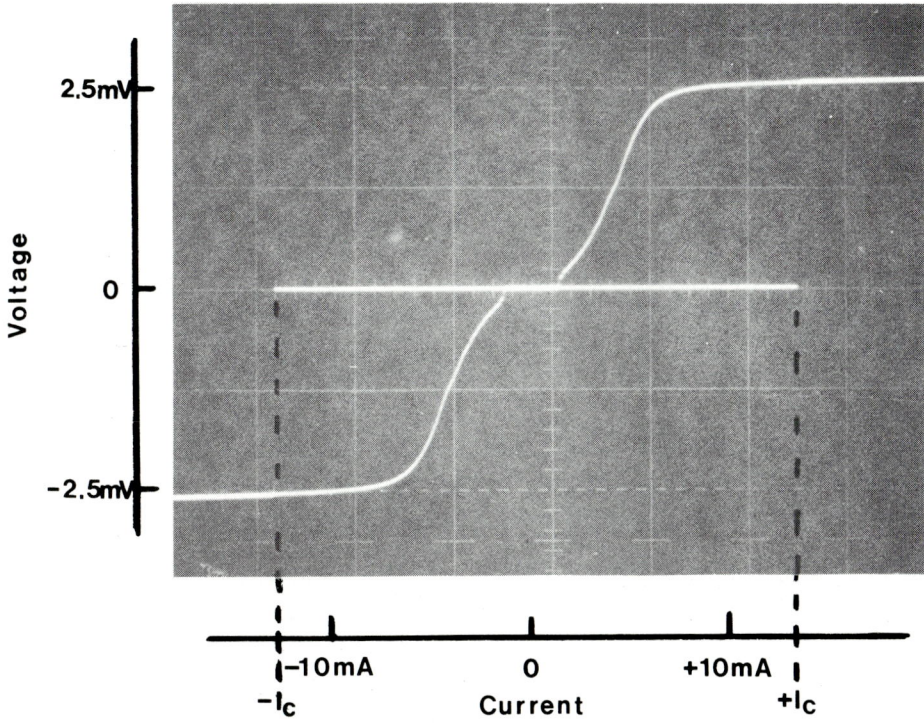

FIGURE 1.   Current/voltage characteristics of a tunnel junction showing the dc Josephson effect.

PLATE II (ROBINSON)

A NOVEL RADAR SET

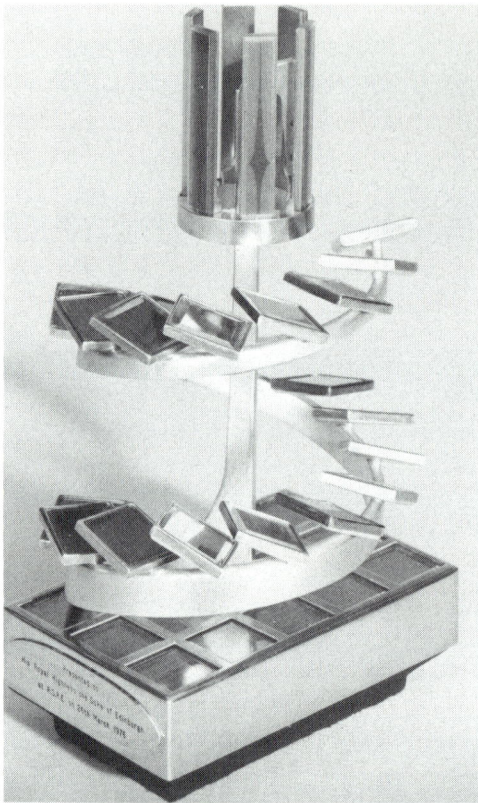

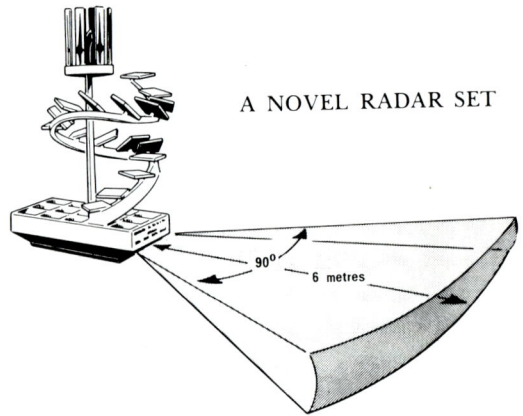

FIGURE 4.   The radar set detects people moving within the indicated volume. It operates on a well lit surface in reasonably strong daylight or artificial light. Sufficient drive comes from the spiral array of solar cells. It can be placed behind a glass door with little change in sensitivity. The visual indicator panels in the form of the crown at the top contain a new stable liquid crystal material.

Figure 12. MADGE station guiding helicopters.

PLATE IV (ROBINSON)

FIGURE 17.   Laser chip on mount.

materials. Around 1970, the Royal Signals and Radar Establishment re-appraised the transferred electron effect and considered that an alternative semiconductor, indium phosphide, would give a considerable improvement in microwave performance. Theoretical studies showed that microwave conversion efficiencies greater than 20% should be possible with an optimized device, more than double that obtained from GaAs. Recent microwave results from indium phosphide at 15GHz have borne out the original predictions and table 1 demonstrates the best combination of peak power, long pulse capability and mean powers with high efficiency that has been obtained from any transferred electron device, or indeed from any other solid state power device at these frequencies. This advance allows development of many new and exciting radar applications.

TABLE 1

Performances for Single MESA Pulsed InP
Transferred Electron Oscillators

Oscillator selected for:	Area (cm^{-2})	Duty Cycle	Peak Power (W)	Mean Power (mW)	Pulse Width (μs)	Efficiency (%)	Frequency (GHz)
Peak power	1.0×10^{-3}	0.001	21	21	1	15	15
Mean power	1.0×10^{-3}	0.01	15.0	1500	1	13	15.9
Efficiency	4.0×10^{-4}	0.001	5	5	0.5	24	12.3
Pulse width	1.0×10^{-3}	0.01	10	100	50	10.5	14

Liquid Crystals—The Biphenyls: A New Family of Liquid Crystals

The early development of liquid crystal displays was delayed because the liquid crystal materials available were unstable and the devices made with them had a short life. In joint research with Hull University during 1972, a new family of very stable liquid crystals, the cyanobiphenyls, and subsequently the cyanoterphenyls were discovered. An example of a biphenyl is:

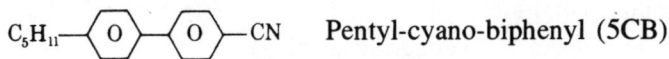

C_5H_{11}—⟨O⟩—⟨O⟩—CN   Pentyl-cyano-biphenyl (5CB)

Single compounds only exhibit liquid crystalline properties over a narrow range of temperature. To obtain liquid crystals with wider temperature ranges for use in displays, eutectic mixtures had to be designed using a complex technique. The eutectic phase diagram of a simple binary mixture is shown in figure 6, but mixtures for use in displays contain many components. For example E7, which is a widely used mixture of biphenyls and terphenyls, is a four-component eutectic with a nematic range of $-10°C$ to $+60°C$. The choice of components for the eutectic mixtures has a great influence on their properties in a display

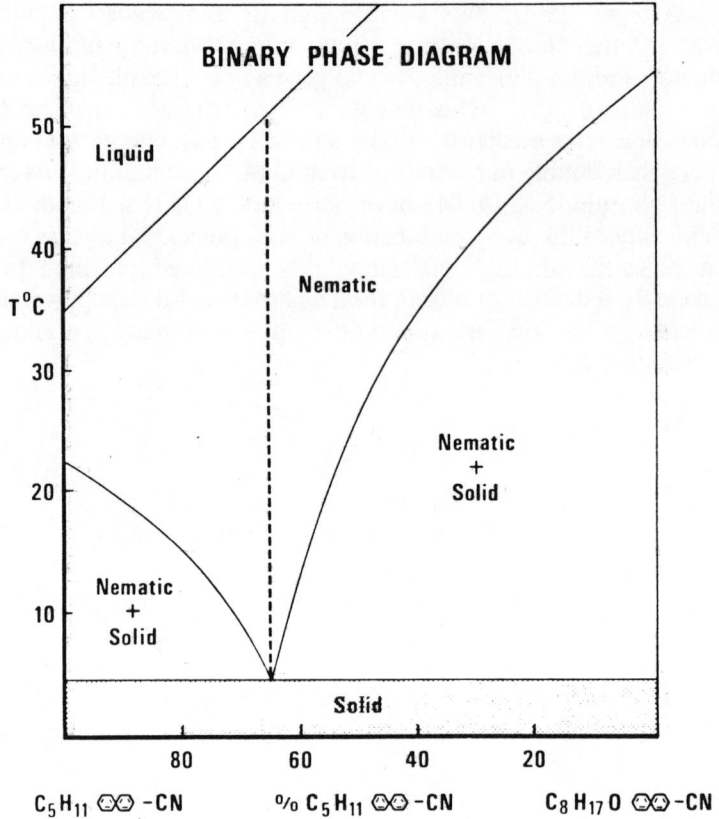

FIGURE 6.   Binary phase diagram of mixtures of two biphenyls.

device, and a considerable effort has been made to optimize the performance for specific applications. In general, however, mixtures of biphenyls possess a unique combination of properties:

(i) chemical and photochemical stability
(ii) low toxicity
(iii) low operating voltage in displays
(iv) quick response times in displays.

BDH Chemicals Ltd are licensed to manufacture and sell biphenyls. Their sales now account for about 50% of the world consumption of liquid crystals, of which the majority is the mixture E7.

Liquid crystals find their main use in electro-optic shutters used as display devices. A 10μm thick layer of the liquid is held in a flat cell made from glass plates with inner surfaces coated with patterned transparent conducting tin oxide layers. These same surfaces are treated to induce the long liquid crystal molecules to lie in a preferred direction, and the directions on the front and back of the cell are made orthogonal so that the optic axis of the liquid layer twists through 90° (figure 7). In figure 7, left, the twisted layer rotates the plane of polarization of transmitted light by 90°. When the cell is placed between crossed polarizers, this rotation allows light to pass through the system. In figure 7, right, the application of about 3 volts between the conductor patterns on either side of the liquid crystal reversably removes the twist and the rotation effect, and light is no longer transmitted. The phenomenon is called the twisted nematic electro-optic effect, and the devices, together with appropriate electrode patterns, can be used as versatile displays. A reflective display for use in ordinary ambient lighting is made by placing a reflector behind the analyser.

Twisted nematic displays have the following advantages:

(i) visible in bright ambient lighting conditions

(ii) low voltage, low power operation (3V, 1μw cm^{-2})

(iii) flat encapsulation, typically a few millimetres in thickness

(iv) simple and cheap to manufacture

(v) versatile format

(vi) long life (>10 years).

Because of the above features, the twisted nematic effect has become well established in the manufacture of digital displays, e.g. for wristwatches and pocket calculators. The twisted nematic displays shown in figure 4 are used as

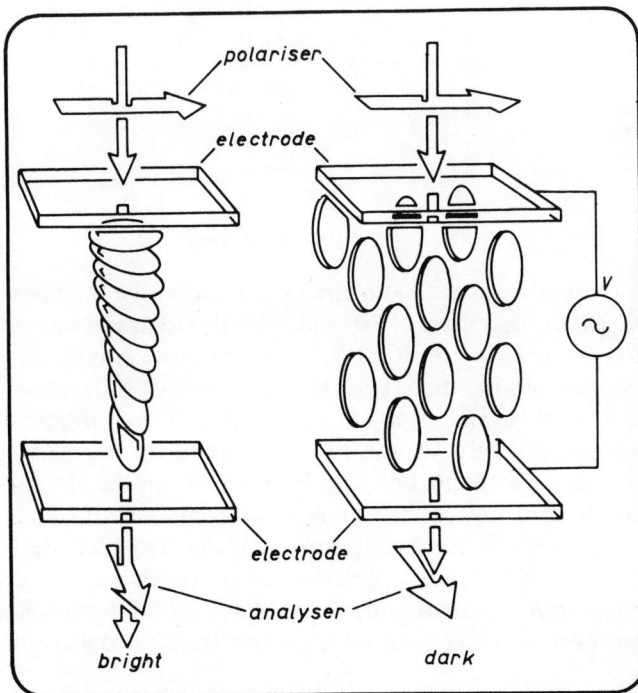

FIGURE 7.   Construction and operation of a twisted nematic cell.

indicators which simply demonstrate their low power and light modulating capacity. No other known display technology could be used for displays of such an area requiring a power input of only a few microwatts.

Microwave Angle Measurement

The wave nature of electromagnetic radiation has been utilized to give precise measurement of distance, velocity and direction. Angle measuring techniques in the microwave radio-frequency band are being used increasingly for navigational purposes. Two related techniques for precision aircraft guidance which are currently under investigation are described.

Figure 8 shows the safe approach path of an aircraft down to a height of about 50 feet above the runway. The aircraft can descend along this safe path with knowledge of its elevation angle relative to the horizon (at E) and its azimuth angle relative to the runway centreline (at A). Radio antenna arrays on the ground provide this angular information and deviations from the safe path are displayed to the pilot. The information enables manual or automatic approaches to be made.

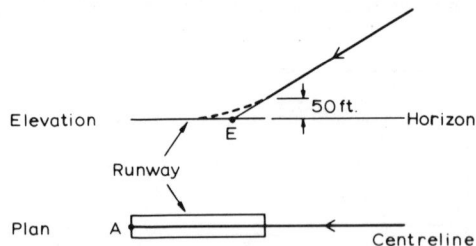

FIGURE 8.   Aircraft approach path.

Guidance is currently obtained at major airports from the Instrument Landing System (ILS), an array operating at 110 and 330 MHz (3 m and 1 m wavelength). Microwave arrays operating at 5000 MHz (6 cm wavelength) are now being developed. They are smaller than ILS arrays and offer significant advantages in performance, flexibility and ease of installation. These improvements are needed to meet increasingly stringent landing guidance requirements. Three types of system are now under investigation. These are the 'Interferometric', which measures the direction of a signal received from an aircraft and returns this information to the aircraft via a data link, and the 'Doppler' and 'Scanning Beam', both of which transmit angularly coded signals to a receiver on the aircraft. The three techniques, although apparently very different, have broadly comparable performance characteristics arising from an underlying physical similarity.

Certain environmental factors are common to the three systems. The choice of the 5000 MHz operating frequency is a compromise; at higher frequencies the

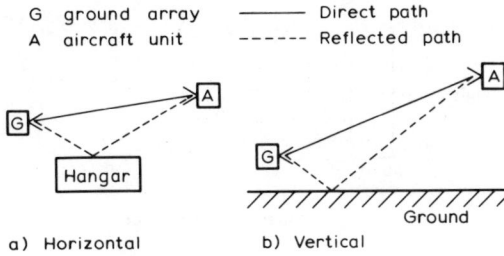

FIGURE 9.   Multipath reflections at an airfield.

attenuation in heavy rain becomes significant, at lower frequencies the array length increases and can become a problem. 'Multipath' or reflected signals interfere with the direct signal and affect the angular accuracy. Figure 9 shows the main sources of such signals at an airfield. Reflections from hangars or taxi-ing aircraft affect the horizontal (azimuth) guidance. Reflection from the ground affects the vertical (elevation) guidance. The system design must therefore take account of multipath.

To illustrate the different approaches the principles of Interferometric and Doppler systems are now described.

Interferometric System

Figure 10 shows an interferometric system known as MADGE (Microwave Aircraft Digital Guidance Equipment). It consists of azimuth and elevation receiving arrays and a ground–air data link. The aircraft interrogates the ground

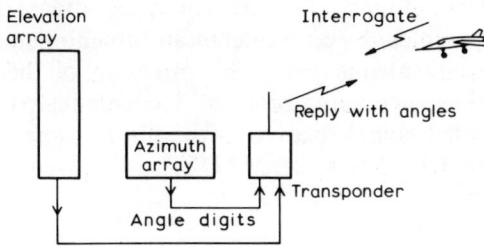

FIGURE 10.   MADGE schematic.

equipment with a coded pulse train; angle information, in digital code, is transmitted back to the aircraft via the data link transponder. Range to touchdown is derived from the go-and-return time of the transmission. Messages are identified by digital codes designating the aircraft and ground stations.

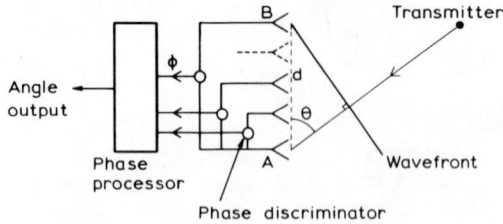

FIGURE 11.   Interferometer principle.

Figure 11 shows the principle of angle measurement. The signal phase is measured at a number of points across the receiving array aperture AB relative to the signal at A. It can readily be shown that the measured phase is related to the signal direction θ as follows:

$$\varphi = \frac{2\pi}{\lambda} \, d \, \cos \theta \qquad (1)$$

where φ is the phase difference between a pair of antennas spaced by a distance d and λ is the signal wavelength.

Thus θ is derived by measurement of φ and knowledge of d and λ.

The azimuth unit is a thinned array in which antenna pair spacings increase in geometric progression (2:1 ratio). The phase at the widest spaced antenna pair AB provides accurate angle measurement. However, it contains unknown integer multiples of 2π giving rise to ambiguity, which is then resolved by reference to the phases at the more closely spaced antenna pairs.

The elevation unit is a filled array where the spacings progressively increase by equal increments. This more complex array is needed because the wavefront distortion arising from ground reflection is more significant in the vertical plane. The wavefront is 'sampled' by each antenna and unambiguous phase differences are derived for each antenna pair. The direction of the wavefront is then determined by a phase averaging process. The antenna patterns are shaped to attenuate the reflected signal relative to the direct signal.

Figure 12, plate III, shows the MADGE ground equipment in a typical helicopter operation.

Doppler System

Figure 13 shows a 'Doppler' transmitting array. Two signals are generated. The first is obtained by switching energy sequentially to a linear array of antennas; the second, an offset reference signal, is radiated from a fixed antenna. The sequentially switched signal can be regarded as originating from a single antenna moving along the array with velocity v. The velocity component in the

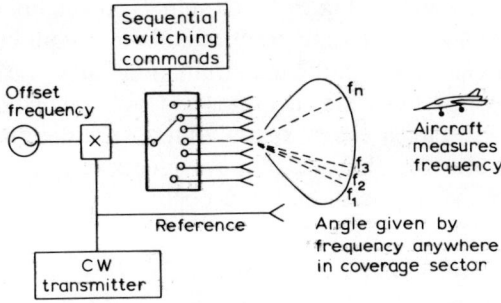

a) Transmitting array

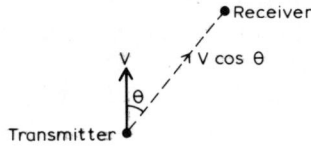

b) Velocity component

FIGURE 13. Doppler ground array.

direction of an airborne receiver subtending an angle θ is $v \cos \theta$. The receiver measures a Doppler shift f_D between the commutated signal and the reference signal where:

$$f_D = \frac{v}{\lambda} \cos \theta. \qquad (2)$$

Thus angles relative to the array are 'labelled' or encoded by a frequency difference.

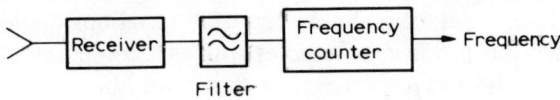

FIGURE 14. Doppler airborne receiver.

The airborne equipment (figure 14) is relatively simple. Following amplification in a receiver the signal is filtered and the frequency is measured. Azimuth and elevation transmissions are time shared. The reference signal is offset by a frequency f_o (typically 100 kHz) relative to the commutated signal so that a frequency $f_o + f_D$ is measured at the receiver. This avoids the need to measure very low frequencies when $\theta \simeq 90°$ and enables signals at positive and negative angles to be distinguished easily. The frequency is measured by a cycle counting technique. In elevation, ground-reflected signals are interpreted by the airborne receiver as negative angles. Such signals are readily removed by a fixed filter, as are other signals coming from outside the required coverage limits.

Further significant multipath discrimination may be obtained with a narrow-band tracking filter which acquires and locks on to the Doppler signal. A recent development makes use of spatial diversity at the ground transmitter to reduce in-beam multipath. Because of the sequential nature of the transmissions this is possible with simplicity and low cost.

Relationship between Doppler and Interferometric Systems

Doppler and interferometric systems can broadly be regarded as reciprocal. Suppose signals of equal frequency were radiated from antennas A and B in figure 11 and received in a direction θ. Equation (1) now defines the phase difference between the received signals. If antenna B is caused to move along the array at velocity v, then $d = vt$ (where t is time) and

$$\varphi = \frac{2\pi}{\lambda} vt \cos \theta. \tag{3}$$

There is now a frequency difference between the signals given by

$$f_D = \frac{1}{2\pi} \frac{d\varphi}{dt} = \frac{v}{\lambda} \cos \theta. \tag{4}$$

This is seen to be the same as relation (2) derived from the 'Doppler' concept.

APPLICATIONS OF PHOTON CORRELATION

In many optical experiments and applications of optical methods information is contained in the temporal variations of detected light. This is particularly true in the field of laser scattering. Such temporal variations could be essentially periodic, as when for instance, a Doppler shift is recorded by scattering a laser beam from a steadily moving object, or they could be essentially random as when a laser beam is scattered, for example, from a polymer solution. Mixtures of these two extremes can also occur. The study of such fluctuations on time scales between, say, 1 and 10^{-8} s has developed into an important new area for optics in the last few years.

There are two basic methods for studying temporal variations of signals of these types, namely, spectral (frequency) analysis and temporal autocorrelation. In the former method the signal is passed through an electrical filter which is tuned successively through a range of frequencies, while in the latter the averaged product of the signal is taken with earlier values of itself generated by passing it through successive fixed time delays. For stationary signals (that is signals whose properties are independent of time origin), it can be easily shown that the spectrum and the autocorrelation function form a Fourier transform pair (Wiener–Khinchine theorem).

The desired information content of the temporal fluctuations can be acquired most efficiently by processing the set of spectral intervals or the set of correlation coefficients in parallel, the first by means of a bank of filters or the second by means of a series of fixed delay circuits. While it is sometimes possible to construct a bank of suitable electrical filters covering adjacent regions of frequency, in general this is much less easily accomplished than is constructing a series of time delays, particularly so nowadays with the availability of cheap integrated circuit shift registers.

Parallel channel autocorrelation is thus an effective way of investigating temporal fluctuations. In the case of optical detection where, after a number of years of research in photon statistics, fast modern photomultiplier tubes and circuits are available to give the signal in the form of an undistorted digital stream of standardized individual photodetection impulses, we have a perfect match with the digital operation of saturated logic semiconductor devices and extremely efficient signal processing is possible.

Such digital, parallel channel autocorrelation of individual photodetections has become known as photon correlation; it provides an optimum method of recovering information from many forms of optical signal in quite an astounding range of scientific fields. These range from the measurement of spermatozoal mobility to noise suppression studies in the Concorde aeroengine. As an example, a photon correlator system measuring blood flow velocities in the retina is described here.

Figure 15 shows the optical pathways of the instrument. At the end of the light path, the photomultiplier receives beat frequencies in the kilocycle range

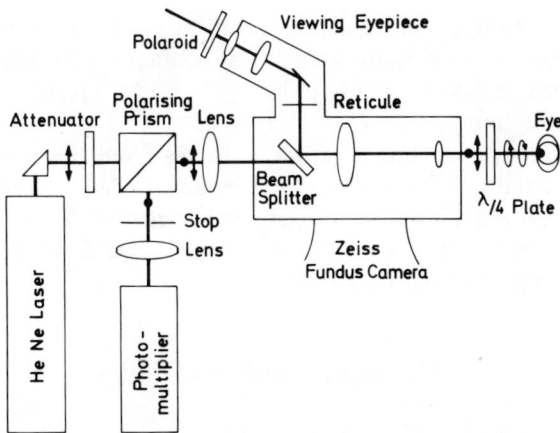

FIGURE 15.   Monostatic heterodyne velocimeter using polarizing optics for retinal blood flow studies.

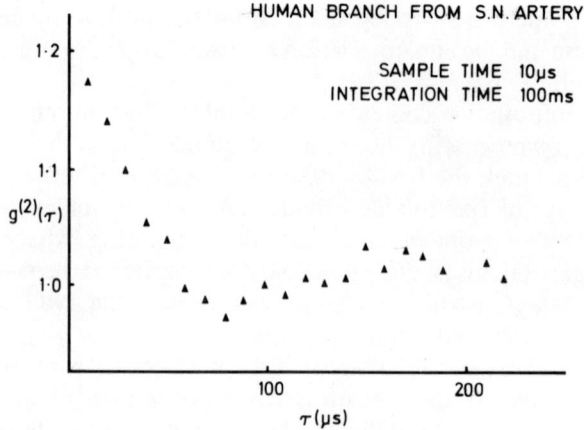

FIGURE 16.   Photon correlation function from human retinal blood flow.

corresponding to each of the red blood cell velocities present in the blood vessel. These homodyne frequencies result from the slight frequency shift of the light scattered by the red cells, as compared with the frequency of light scattered by the adjacent vessel wall. The homodyne signal is at individual photon level, randomly occurring with a compound Poisson distribution; consequently the signal is only distinguished by electronic correlation of consecutive samples collected over a period of time. The occurrence of events in successive samples is correlated with previous samples up to the limit of the integrating time. Figure 16 shows a photon correlation function from a human eye. The slope of the first descending limb, being chiefly influenced by the faster velocity components in the centre of the vessel, is directly related to the maximum velocity of blood flow.

In the most recent trial, reproducible recordings were made in man using only 1 μW laser power. At the same time the anaesthetized cat was studied, using 9 μW for 445 ms to cover a whole pulse cycle, and the results checked by ciné fluorescein angiography.

Work is in progress to translate these studies from the feasibility stage to clinical instrumentation, the low level of laser beam intensity and short sampling times suggest that a useful clinical tool can be created. Since the technique is entirely non-invasive, repeated measurements should be possible, greatly widening the scope of investigations.

COMMUNICATION BY LIGHT

Communication by light is nothing new—witness the use of mirrors by the British Army on the North West Frontier and the Navy's Aldis lamp. However, these are slow and confined to single messages at a time. Anyway why use light at all—what's wrong with electrical systems like radio or the present telephone? The answer is one familiar in other aspects of today's life—overcrowding. Radio stations are overlapping (listen to the medium wave band on your wireless set one evening) and although by using modern technology we can cram many telephone conversations along one copper conductor, the underground ducts containing the copper cables are full to bursting. Some new medium is needed: we find it in the world of optics.

Communication Principle

Light is electromagnetic in nature but lies further along the spectrum than radio, and, because of this, can carry much more information, as much as 100 000 times more. This fact has been known for many years, but it was only twelve years ago that two Standard Telecommunication Laboratories scientists, Dr. Charles Kao and Dr. George Hockham, identified the requirements. They proposed that glass fibres carrying light signals from lasers could be used. All that was needed was a laser, small, powerful and with a long life, a hair-fine glass fibre of kilometre lengths and a device to detect the light after transmission. The trouble was that none of these existed! However, they postulated that the burgeoning technology of semiconductor physics could provide both solid-state lasers and detectors and the method of producing glass of sufficient purity. They were right, but it has taken twelve years of intensive research to prove it.

Method of Transmission

To appreciate the progress which has been made and the promise for the future, it is necessary to understand a little of how optical communications works. The initial signal, voice or data, is first converted into a pulsed electrical current. This current is applied to a solid-state light source such as a light-emitting diode (similar to the tiny red lamps which make the numbers in pocket calculators) or its more powerful relative, the semiconductor laser. This source is excited by the current and emits a bright, near infrared, light which pulses in exact proportion to the modulations of the current by the signal. The light output is transmitted along a fibre, detected, and decoded at the receiving end into the original signal.

One of the important properties of laser-emitted light is its ability to maintain its power over great distances, with practically no scattering. The 'spot on the moon' was an early demonstration of this power. The stumbling block seemed to be the lack of means to guide the light and to protect it from accidental scattering by solid objects or atmospheric particles. A 'wave guide' or 'light pipe' was an obvious answer, but the development of a suitable one which could preserve all the properties of the light was another matter. Glass fibres were examined, of course, along with other methods, but the weakness at that time was the inability of the glass to contain a decent signal for more than a few metres. To achieve the kind of clarity needed for a useful system, you need a glass from which could be made a window 2 kilometres thick, through which you could see perfectly! What made Kao and Hockham's idea so important was that they made measurements which suggested that it would be possible to eliminate most of the impurities which caused the light loss in glass. Although working with glass fibres having losses of thousands of decibels per kilometre, they predicted glass fibres could be developed with losses of no more than twenty decibels per kilometre. On that assumption, deducing also that the laser as a light source would become a practical device, they made their forecast for optical telecommunications.

The Glass Fibre

Now, how does the light travel along the fibre? The significant words are 'total internal reflection', which some may remember from school physics. When light strikes a surface it may pass through the surface or may be partly or completely reflected by it. Because of this, if you look down a rod of glass the inner surface appears to be a mirror. Light will reflect off this mirror surface and thus travel along the rod. This is an example of total internal reflection. Glass fibre used in optical transmission is made from two glasses, one inside the other. The light is contained in a zigzag bounce along the fibre, reflecting off the optical 'wall' between the core glass and the cladding glass. This rather complex construction is achieved by pulling the fibre from a molten glass rod which is in itself composed of the two glasses moulded together. The technique is remarkably similar to that used in the manufacture of seaside rock when the sweet is drawn down to size with the lettering running through it. As all of the predicted improvements in the purity of the glass have been achieved, it is now possible to produce signal losses as little as two or three decibels per kilometre. This means that our window is now some 16 kilometres thick or as clear as fresh air on a bright day.

Input and Output Devices

The semiconductor laser we are talking about is very different from the gas laser used to cut steel or concrete. It is literally grown as a crystal in compounds of the metal gallium, which is then divided into tiny chips, each being not unlike a multi-layer sandwich. When the pulsed electric current from the signal is passed into the laser its effect is compressed into the central layer, where light is generated and emitted, confined in an intense, near parallel beam. Lasers of this type now have lives which can be measured in many thousands of hours and by calculating the rate of change it is possible to predict that they can achieve the very long operating life necessary for use in practical telecommunications systems.

In use the laser chip is butted very precisely against the end of the fibre. Once again an appreciation of scale must be made; the single laser chip is barely visible without a microscope; for handling purposes it must be mounted on a brass plug to which the fibre is attached (figure 17, plate IV). In the ultimate system the laser would emit its light into a fibre with a core diameter of only 3 micrometres, that is three thousandths of a millimetre. As 3 micrometres is the wavelength of the near infrared light emitted, this means that a single perfect light wave could be launched into the fibre. Such a pure signal could carry 150 000 telephone calls or their equivalent! The practical systems now coming into use do not have this ultimate capacity, but nevertheless can carry several thousand voice channels down each of eight optical fibres all contained within a cable no bigger than an ordinary domestic 13 amp lead.

Semiconductor research has also enabled new detectors of light called avalanche diodes to be built. These are capable of detecting very, very low light levels without introducing noise. They enable the signals emerging from the fibre to be converted back into the electrical impulses necessary for driving the standard telephone apparatus.

The System

Research has also been carried out in connection with the mechanics of the complete system, that is, development of joints, splices and connectors to enable it to be assembled in the real world of roadside manholes and telephone exchanges. Tests have been carried out which have shown that the optical cable is far stronger than the equivalent copper cable, a paradox indeed considering the normal brittle behaviour of glass. Its extremely small size enables it to be pulled into existing cable ducts where only little space remains between existing cables, or alternatively, if the same duct is filled to capacity with optical cables, its capability is many thousand times that when it contained only copper. Other advantages accrue from the fact that the light is completely inert. In other words there are no electrical sparks to cause danger in areas of possible gas concentration such as mines; also there is no radiation from the cable and therefore it is practically undetectable and is a very secure means of transmitting information. As the fibre is made from some of the world's most common raw materials—sand, silica—optical cable would free the diminishing copper resources of the world for other uses and conservation; it is also inherently less costly. Not that very large quantities of the material would be required; it has been estimated that a cable television system to supply over 5 million homes in the UK requiring a total of 40 000 000 kilometres of cable could be produced by having just 800 metric tonnes of pure glass available.

Conclusion

Optical communications is an infant technology. In its so far eleven-years life it has shown that it can be used in practical systems. There are considerable areas as yet unexplored in which other uses can be exploited; for instance it may be possible to work entirely in the optical regime with no electric currents flowing in the system at all.

CEEFAX—THE BBC'S TELETEXT SERVICE

The provision of a broadcasting service of 'written' information had long been recognized as having great potential but it did not become practicable until the advent of advanced solid-state technology. CEEFAX makes available about two hundred different 'pages' of news and topical information on a wide range of subjects and is transmitted along with the BBC-1 and BBC-2 programme services. A viewer whose television receiver is equipped with the necessary Teletext decoder can switch to any of these pages at any time when the programmes are on. The selected page is displayed on the television screen instead of the programme picture (or superimposed on it) for as long as the viewer wants and he is free to switch to another page or revert to the programme at any time.

CEEFAX could be used to provide subtitles to help deaf people to enjoy television and ways of doing this are being studied. The transmitted signal could describe perhaps one or two lines of characters superimposed on the television picture by circuits within the receiver and only the viewer who wishes to use the facility will see the subtitles. News flashes can be handled somewhat similarly and if a viewer selects the appropriate page he can watch the programmes normally in the knowledge that, if a news flash is transmitted, it will automatically be superimposed on the programme picture.

Television receivers incorporating Teletext decoders are now becoming widely available and there is great public interest in the service.

The Basic Data Channel

CEEFAX uses an existing communication channel, the network of 625-line BBC-1 and BBC-2 television transmitters. Like most communication channels, television has some spare capacity and part of this is used to carry the CEEFAX signals which thus ride 'pick-a-back' on the television picture signals. It would not be possible to take advantage of this spare capacity were it not for the advances in technology that have taken place in recent years in the development of large scale integral (LSI) circuits.

How CEEFAX is transmitted

Many different ideas were considered to take advantage of unused frequencies within the television video bandwidth or unexploited time during transmission and it was agreed that the most promising approach was to make use of some of the unused lines in the field blanking or vertical retrace period. (It has been common practice for some years for broadcasters to make use of lines during field blanking for their own test and control purposes.)

So, two lines (data-lines) during field blanking are used to carry CEEFAX pulse signals. In the receiver, the data pulses are 'written into' a store which accumulates the data for a complete page. Although the data-rate during the data-lines is very high, only two data-lines per field are used and the average data-rate is relatively low. In the receiver the accumulation of data for one complete page takes just under a quarter of a second. The stored data signals control special alphanumeric character-generating circuits which produce video signals describing the required characters for the television screen. The transmitted signal is thus quite unlike an ordinary television signal, which directly represents the image required but is more like a telegraph signal which controls the operation of a local teleprinter according to a pre-arranged code. The process of 'reading' the stored data and displaying the generated characters on the television screen takes place repeatedly using normal television scanning.

Transmission Sequence

The CEEFAX editorial staff generate the coded signals describing each page from the keyboard of a video display unit. The complete set of coded signals representing all the 'pages' in a 'magazine' is then assembled in the store of a specially programmed computer system. The contents of the store are next transferred, in 'packets', to the allocated lines in the broadcast television signal. All the packets describing one page are transferred in succession, the packets describing the next page following, in sequence, until all pages have been dealt with; the cycle then starts again with the first page.

At the receiver, the viewer's choice of page to read is made by operating buttons on a keypad and this causes the incoming data-signal packets describing the required page to be loaded into the receiver's page store, whence they are 'read-out' to provide continuous control signals for the character generator. The viewer's action also causes the output of the character generator to be superimposed upon or to replace the displayed television programme picture.

A few seconds may elapse before the selected page is displayed at the receiver. This is because of the cycle of operations in which all pages of the whole magazine are transmitted over and over again. Each page is formed from up to 24 rows and, with 2 rows transmitted per field, 0.24 second is required to transmit a complete page. The magazine has a capacity of 100 pages and the transmission in sequence of a complete magazine would therefore take 24 seconds and in these circumstances, at worst, a viewer may have to wait as long as this before he sees the page selected. In practice, however, two factors combine to reduce the waiting time. First, it is unlikely that a complete magazine will be transmitted and, secondly, the use of a method of working which ensures that all-blank rows are not transmitted at all reduces the transmission time for the whole magazine to about 12 seconds typically, with an average waiting time of about 6 seconds. This point is probably a very important factor in determining the public acceptability of CEEFAX.

The above description has referred only to pages containing alphanumeric characters but graphics displays are possible by using 'filled-in' (or blank) character rectangles assembled to form the pattern required. Also no mention has been made of colour which is available at the editor's option in both alphanumeric and graphics displays.

DEVELOPMENTS IN ELECTRON MICROSCOPY AND MICROANALYSIS

M.L. JENKINS AND M.J. WHELAN, F.R.S. (Editors)

Department of Metallurgy and Science of Materials,
University of Oxford

This chapter was prepared from contributions submitted by the following: G.R. Booker, V.E. Cosslett, F.R.S., P. Duncumb, F.R.S., A.M. Glauert, M.J. Goringe, B.L. Gupta, R.W. Horne, M.H. Jacobs, M.L. Jenkins, T. Mulvey, W.C. Nixon, K.C.A. Smith, P.N.T. Unwin, M.J. Whelan, F.R.S.

1. INTRODUCTION

The past 25 years have witnessed a remarkable growth in the availability of electron microscopical and microanalytical instruments, and in their application in most fields of science and technology. The extent of this growth can be appreciated by noting that in 1952 metals could only be examined in the electron microscope by means of surface replicas, thin sectioning techniques for biological tissues had just been introduced, the direct observation of crystal lattices and of dislocations in them was still four years away, the first scanning electron microscope was still under construction and the scanning microprobe analyser existed only on paper. In that year there were only 80 electron microscopes in the whole country; of these only a minority could operate at 100 kV. Now there are over a thousand, including nine very high voltage instruments.

The development of new instruments has gone hand in hand with the evolution of preparative and observational methods needed for their successful application. We can now cut sections 500 Å thick and prepare equally thin metal specimens as a matter of course, examine their structure and composition at any voltage from 10 kV to 1 MV and their surface features in a scanning microscope. The heavier atoms can already be imaged individually and lattice planes with a spacing of less than 1 Å can be resolved. We now look forward to a true point-to-point resolution of better than 2 Å compared with the 20 Å we were glad to get in 1952.

Historians of technology should note that nearly all these innovations originated in academic laboratories and were primarily stimulated by a need to meet particular demands in research. Of course, once the value of a new instrument had been established, industry became willing to take it up. British firms have been making a practical contribution to electron beam

137

instrumentation equal to that of any other country, especially in respect of the microprobe analyser and the scanning and high voltage microscopes. Many laboratories in this country have made equally important advances in the applications of these new tools of research. The display at the Royal Society's Jubilee Exhibition and the articles in this chapter illustrate these developments, portraying the present state of the subject and taking a look into the future.

2. Developments in Electron Optics and Electron Optical Instrumentation in the UK

2.1. *Historical*

Electron microscopy as we know it today began effectively in 1933 when E. Ruska at the Technological University of Berlin succeeded in building an experimental electron microscope that surpassed the optical microscope in resolving power. At first, support for a commercial development was slow in forthcoming. However, in 1936 the Metropolitan Vickers Electrical Company in Manchester was approached by Professor L.C. Martin of the Imperial College of Science and Technology with a request that they design and construct an electron microscope for the purposes of evaluating critically this new method of microscopy. This project was greatly helped by the good offices of The Royal Society who provided some funds towards the cost of manufacture. The new microscope, the EM1, was installed at Imperial College towards the end of 1936 and provided valuable operating experience, as well as stimulating commercial development in Europe and the United States. Unfortunately, Metropolitan Vickers were unable to spare further engineering effort to put the electron microscope into production and the project was shelved. After 1945 development proceeded rapidly, and the first British serially produced electron microscope, the EM2, was installed at the Dunlop Rubber Company in Birmingham in December 1946.

The Jubilee period (1952–1977) that we are now celebrating coincides very happily with a memorable contribution to electron microscopy by the scientists and engineers of this country. Thus in 1952 a simple electron microscope, the EM4, of ingenious design had been produced for routine laboratory use; the first production instrument was delivered in June 1953 to the French Military Attaché. The EM4 had a guaranteed resolving power of 100 Å, then considered very adequate for most practical applications of electron microscopy. In a parallel development, the EM3 electron microscope with a guaranteed resolving power of 25 Å was designed and manufactured in the early 1950's. In May 1958 the first production model of a new high resolution electron microscope, the EM6, with a resolving power of 10 Å and a maximum magnification of 120 000 was installed at the Royal College of Surgeons by Associated Electrical Industries (AEI) Ltd, into which group the Metropolitan Vickers Electrical Company had been incorporated. This was followed by the EM6B high resolution biological electron microscope with a maximum magnification of 250 000 and a resolution of 5 Å. These instruments, especially the last two, paved the way for further developments such as the analytical electron microscope and the high voltage electron microscope described below. More recently, the manufacture of scanning electron microscopes and microanalysers by the Cambridge Scientific Instrument Company, and of scanning transmission electron microscopes by AEI Ltd and Vacuum Generators Ltd has made available instruments which have proved exceedingly useful in the fields of materials analysis, molecular biology, medicine and micro-electronics, to name only a few of the disciplines which have benefitted from the science of electron optics.

2.2. Developments at Cambridge University

After the end of the 1939–1945 War Cambridge became a leading source of advances in electron optics and related instrumentation. At the Engineering Department, under C.W. Oatley, the scanning electron microscope was developed into a practical tool with wide applications. From the Cavendish Laboratory, under V.E. Cosslett, came the point-projection X-ray microscope, the scanning X-ray microanalyser and the first high-voltage microscope in the country, all of which found their way into commercial production. More recently the two laboratories have united their efforts in designing and constructing a high resolution electron microscope.

The Cavendish Laboratory. In 1946 work in electron optics was begun and proceeded in two directions: the production of point sources of X-rays and the study of electron lens aberrations. Earlier experiments on focusing electrons into a fine probe, carried out at the suggestion of J.D. Bernal, FRS, had impressed Cosslett with the potentialities of a point X-ray source but also depressed him by the magnitude of the errors of magnetic lenses. Following up the theoretical studies of O. Scherzer on chromatic and spherical aberration, a search was made for a practical correction system. A thorough analysis by P.A. Sturrock was the first outcome. Later P.W. Hawkes and his collaborators extended the mathematical treatment and measured the electron optical properties of sets of quadrupole and octopole lenses. Although partial success was achieved in correcting spherical aberration, no readily adjustable corrector could be devised. Later studies of pulsed lenses offered little greater prospect of success. Attention has since been concentrated on image processing for deconvoluting the effects of spherical aberration and defocus from a micrograph.

Existing electron lenses, however, are capable of focusing a beam to a small probe, less than a micrometre ($1\ \mu$m) in diameter, which then generates X-rays in the target or specimen on which it falls. If this target is a metal foil the X-radiation emerges from the far side as from a point source and can be used to form X-ray images by geometrical projection. After preliminary experiments by D.A. Taylor, the point-projection microscope was made into a working instrument by W.C. Nixon as a Ph.D. project. His original (1952) model was shown in the Jubilee Exhibition, together with examples of its applications in angiography and materials testing. It readily gives an image resolution better than 1 μm and with a very thin target can approach 0.1 μm, but beyond that it is limited by diffraction and mechanical effects, since the X-ray intensity falls and exposure times become unduly long. Apart from direct imaging the point-source can also be used for local analysis of specimens by absorption or fluorescence methods.

In 1952 an enquiry from the MRC Pneumoconiosis Unit at Cardiff, about means of identifying silica in miners' lungs, prompted the design and construction of an instrument for microanalysis by X-ray emission spectrometry. Based on Castaing's original static microprobe (with electrostatic lenses), this used magnetic lenses and included a scanning system to make visible the area of the specimen being analysed, an X-ray or back-scattered electron image being displayed on a monitor tube. It was constructed on the carcass of an old RCA microscope by P. Duncumb, as a Ph.D. project, and immediately found applications in metallurgy and mineralogy. Together with D.A. Melford he re-engineered it at the Tube Investments Research Laboratories as noted in § 2.3. In parallel with this work, J.V.P. Long was pursuing the mineralogical applications of the technique with another instrument designed for the purpose. It is not too much to say that the X-ray microprobe analyser, in its various forms, opened a new era in the study of the microstructure and composition of metals, alloys, minerals and biological tissues that contain inorganic elements.

This instrument was followed by the high-voltage microscope, the impetus for which came from metallurgists. From the mid-1950's great advances were made in the electron microscopy of metals, prompted by the discovery of dislocations and the need to study their relation to the macroscopic properties of materials (see § 3.1). The Cavendish Laboratory took a leading part in these developments under P.B. Hirsch, as did Tube Investments Research Laboratories under J.W. Menter. They stimulated the setting up of a committee by the Department of Scientific and Industrial Research to look into the case for a high-voltage microscope (HVEM) in this country, in the light of work with such instruments in France and Japan. The Committee, with Cosslett as chairman, produced in 1961 a fully documented report recommending that the design and construction of a HVEM should be undertaken forthwith. Since no commercial firm in the UK was willing to undertake the task, application was made by Cosslett to the Paul Instrument Fund of the Royal Society for a grant to build a 750 kV microscope at the Cavendish Laboratory, and the Fund provided the greater part of the considerable sum required (approaching £100 000).

With K.C.A. Smith in charge of the project, the Cavendish HVEM was completed by 1966. It was at once seized upon by metallurgists interested in radiation damage problems at AERE (Harwell) and elsewhere, being fully booked for seven days a week for several years until other high voltage instruments became available. These were provided when AEI Ltd put a 1 MV microscope (EM7) into production based on the experience gained at the

Cavendish Laboratory. At present six EM7 instruments are in operation in England, and others have been exported to the USA, France and Germany. Some of the applications of the HVEM were displayed in the Jubilee Exhibition (see § 3.2).

More recently a 600 kV high resolution electron microscope has been constructed at the Cavendish Laboratory in a joint project with the Engineering Department as mentioned below.

The Engineering Department. Research on scanning electron microscopy under the direction of C.W. Oatley began in 1948 and by 1952 a working instrument had been developed by D. McMullen as a Ph.D. project. Improvements were made by K.C.A. Smith during the years 1952–1956 and a wide range of specimens examined including wood and paper fibres. Sufficient interest was created by these results to produce a third generation instrument, designed and constructed by K.C.A. Smith, which was installed in the Pulp and Paper Research Institute of Canada in 1958. This scanning electron microscope was the first and only successful instrument in North America until 1965. It produced some 25 000 micrographs during 1958–1968 and is now in the Science Museum in Ottawa. Interest was slower to develop in this country although J. Sikorski saw the potential of the scanning electron microscope for research in the textile industry as early as 1958, and pointed out its advantages over the replica technique. Applications in many other areas, including the first demonstration of voltage contrast from semiconductors by T.E. Everhart *et al.*, finally led to the commercial production of a scanning electron microscope based on the experience gained in the Engineering Department. The technology transfer was carried out by A.D.G. Stewart who worked as a research student at the Engineering Department during 1958–1962. Stewart joined the Cambridge Scientific Instrument Company in 1962, and the 'Stereoscan' scanning electron microscope was described by Stewart and Snelling in 1964. A great expansion in the use of the scanning electron microscope followed in all the fields where the light microscope had previously been used to image the surface of solid specimens. Higher resolution using a 50 Å electron probe and wider applications led to world-wide manufacture and use of the scanning electron microscope. Examples of micrographs obtained with various instruments are shown in figures 8 and 9 (plates VI and VII).

The same methods of advanced electron optical engineering have been applied to electron beam microfabrication or lithography by students such as A.N. Broers, T.H.P. Chang, J.P. Ballantyne, C. Dix and others under the direction of W.C. Nixon since 1959. A commercial instrument based on this principle for the microcircuit industry has been made by the Cambridge Scientific Instrument Company. Early results of Chang are shown in figure 10 (plate VII). More recently electron beam lithography has been further developed by H. Ahmed using a lanthanum hexaboride gun.

The work has been greatly aided by computer design and analysis of electron optical systems, and computers have also been used for on-line image processing by B.M. Unitt and K.C.A. Smith.

Following on the success of the Engineering Department in electron optical design, construction and operation of the scanning electron microscope, and the similar experience of the Cavendish Laboratory with the 750 kV microscope (see above), the Science Research Council invited the two departments in 1971 to submit a joint project. It was decided to develop an electron microscope for ultra-high resolution, approaching 1 Å, by operating at 600 kV with suitable electron optics and the highest electrical, mechanical and thermal stability. This instrument is now nearing completion in the Old Cavendish Laboratory.

2.3. *Developments at Tube Investments Research Laboratories*

Following the successful development of the X-ray scanning microanalyser in the Cavendish Laboratory in the mid-1950's, a number of applications of the instrument were made, one of the most successful being carried out in conjunction with Tube Investments Research Laboratories. This concerned the development of surface cracks in tube steel which occurred in certain hot-working conditions (see § 3.4). The commercial impact of the work was such that TIRL decided to build an instrument for metallurgical applications, and this was completed by P. Duncumb and D.A. Melford in 1959. Shortly afterwards, this came to the attention of the Cambridge Scientific Instrument Company, who, using the TIRL instrument as a prototype, manufactured and sold some 80 production instruments under the name of 'Microscan' in the following four years. Following further pioneer work by J.V.P. Long in the Department of Mineralogy and Petrology at Cambridge, the Cambridge Instrument Company developed a second generation instrument known as the 'Geoscan'.

During this period, work had started at TIRL on a new instrument, the combined electron microscope and microanalyser, known as EMMA1. The purpose of this work was to develop the facility for studying fine precipitation in metals, by the use of extraction replica and thin film samples, and the successful application of this was reported in 1964. The technique relies upon the greatly reduced electron scattering occurring in thin specimens such as are used in the electron microscope, which permits a much improved resolution for microanalysis over the normal bulk samples; in this instrument an analytical resolution of 0.1–0.2 μm was achieved. Coupled with the analytical facility, the operator can also carry out electron diffraction studies on the area under the probe, thus providing information on crystal structure as well as on chemical composition.

Later, in 1968, work started on a production version, EMMA3, in conjunction with AEI, later sold commercially as EMMA4. The chief characteristics of this instrument were that it was usable as a normal 100 kV transmission electron microscope with an image resolution of 10 Å, and an analytical resolution of about 1000 Å. This was achieved through the use of a new kind of miniature lens, adapted from that developed by J.B. Le Poole in Holland, to focus the probe on the sample, enabling the X-rays to be collected at 45° to the surface in crystal spectrometers of high efficiency. The mini-lens and spectrometers were built into the AEI EM802 electron microscope, a derivative of the EM6 mentioned in § 2.1. Later an energy dispersive analyser was added and the prototype instrument EMMA3 has been engaged in application work continuously since 1970.

3. APPLICATIONS

3.1. *Transmission Electron Microscopy*

Metallurgy and materials

Until the mid 1950's the main method of studying metals in the transmission electron microscope was by surface examination by means of replica techniques, for example the study of surface steps or slip lines after plastic deformation, or the extraction by replica films of features close to the surface such as precipitates. Despite pioneering work by Heidenreich in the USA in the late 1940's, little interest developed for studying thin sections of metals directly by transmission of electrons in the electron microscope. It seems as though the very success of the replica method had drawn attention away from the method of direct examination of thin metal sections. The metallurgist had simply extended to the domain of the electron microscope the methods of studying solid surfaces with which he was already familiar in his use of the reflection optical microscope. In contrast the biologist, who was used to examining thin sections of biological material in transmission in the light microscope, developed more refined methods of sectioning for the electron microscope.

For metals and other crystalline materials, special methods of preparing thin sections of the required thickness are necessary. Since dislocations are introduced by mechanical cutting processes, and since the object is often to study such defects after mechanical deformation in the bulk state, the sections cannot be prepared by cutting with a microtome. Several methods for preparing thin sections are available depending on the particular problem. For bulk specimens, the first stage is usually to cut a thick slice by spark erosion or by an acid-string saw. This avoids appreciable mechanical deformation. The slice is then thinned further by chemical or electropolishing methods to produce thin sections about 500 Å to 3000 Å thick typically. Thin specimens may also be prepared by chemical or electrodeposition from solution or by vacuum evaporation on to a substrate followed by stripping. Ion bombardment thinning (sputtering) is also useful, particularly for minerals or specimens of low electrical conductivity.

In 1956 Hirsch, Horne and Whelan and independently Bollmann showed that dislocation lines and stacking faults could be observed directly in thin foils examined in the transmission electron microscope and thus opened up a whole new field of application of the transmission electron microscope which has burgeoned in the past 20 years. Such lattice defects are visible because their displacement fields perturb the local electron diffraction conditions. This is known as diffraction contrast and has been studied theoretically in great detail.

Figure 1, plate I, is a typical example of a transmission electron micrograph of a thin foil of stainless steel, showing a twin boundary C. Dislocations can be seen on either side of the twin boundary. These appear as short lines which start and finish on opposite surfaces of the foil. They appear curved because of pinning at the surfaces and the occurrence of stresses in the foil induced by the electron beam. Because of these stresses dislocations may be observed to move and interesting studies have been made using ciné techniques of the motion of dislocations in thin foils. At A in figure 1 a dislocation is observed to be dissociated into partial dislocations separated by a strip of stacking fault, and a wide stacking fault traversing the whole width of the grain is observed at B. The stacking fault is made visible by fringes running parallel to the surface intersection. These fringes arise through an electron wave interference phenomenon at the discontinuity in the crystalline arrangement caused by the stacking fault.

Figure 2, plate II, is an example of a technique developed more recently by Cockayne, Ray and Whelan for observing dislocations at higher resolution than was possible previously. The technique uses a dark-field image formed by a weak Bragg reflection, and is thus referred to as the weak-beam technique. For such an image it is only in the regions very close to the dislocation core that the bending of lattice planes is large enough to give contrast. Hence the weak-beam image is very narrow, as predicted by the kinematical diffraction theory of Hirsch, Howie and Whelan. Figures 2 (a) and (b) show respectively strong-beam and weak-beam dark-field images taken in the 220 Bragg reflection of silicon. The increased resolution of dislocation detail in the dark-field image is clearly visible. The weak-beam technique has been used extensively in recent years to study defect geometry at high resolution, the dissociation of dislocations into partial dislocations and the consequent determination of stacking fault energies, and the behaviour of small point-defect clusters produced by radiation damage.

An alternative method of revealing dislocations in crystal lattices was demonstrated by J.W. Menter in 1956. The high resolution of the transmission electron microscope then becoming available was used to resolve the lattice planes in crystals with large lattice plane spacings. A dislocation is then revealed as a terminating lattice plane in the image. The method was first used by Menter for platinum phthalocyanine, where the $20\bar{1}$ planes have a spacing of 12 Å (see figure 3, plate I). However, with the high resolution now available in modern transmission electron microscopes, it is possible to examine terminating lattice planes in metals such as gold and aluminium. Recently the method of resolving the lattice structure has also proved extremely useful in studying crystallographic defects in complex oxides in inorganic systems and is showing much promise as a tool for direct crystal structure determination.

At a somewhat lower level of resolution, transmission electron microscopy has been extremely useful in revealing the arrangement of antiphase domain boundaries which can occur in ordered alloys. The classic example is the copper–gold system CuAuII, where the antiphase boundaries are regularly spaced 20 Å apart, as studied by D.W. Pashley and A.E.B. Presland.

Biology

The growth of applications of transmission electron microscopy in biology has been no less explosive than in metallurgy and materials science. A typical field of application is the study of viruses. Viruses were among the first biological specimens to be examined in the early experimental electron microscopes which were constructed between 1935 and 1940. The study of virus morphology has subsequently followed a parallel course to the development of the electron microscope and specimen preparation techniques. Viruses fall within a size range of about 250–3500 Å and possess well defined geometrical forms. For these reasons they are ideally suited to high resolution electron microscopy which can be carried out in close parallel with physico-chemical techniques (e.g. cell culture systems, immunology, ultracentrifugation, biochemical analysis and X-ray diffraction).

The adenoviruses provide an excellent example for illustrating the progress made in determining the structure and chemical composition of certain viruses. Adenoviruses were first isolated from human adenoid tissue in 1953, and were subsequently found in a wide range of animal hosts. When first photographed in the electron microscope from thin sections of infected tissue, they were seen to be about 700–800 Å across and of angular shape. Many infected cells showed extensive crystalline arrays of adenovirus particles. Much of the cytological work was carried out between 1954 and 1962. With the development of new techniques for isolating and concentrating viruses, it became possible to image individual adenoviruses at higher resolution by directly mounting the particles from liquid suspensions on to specimen supports. Between 1957 and 1962 the experiments showed the adenovirus to be of hexagonal shape corresponding to a polyhedral body, but the details from shadowed and positively stained preparations did not allow the shape and symmetry to be determined with any degree of accuracy.

In 1959, the negative staining technique was applied to highly purified preparations of adenovirus with considerable success. The structure of the protein virus shell was clearly observed as being icosahedral, composed of 252 protein surface units of about 80 Å diameter arranged in accordance with 5.3.2 symmetry. Further work between 1960 and 1965 established that in addition to the 252 surface units, the adenovirus particle possessed 12 spike-like projections from the 12 points of 5-fold symmetry associated with the icosahedral shell. It was also possible to disrupt the virus shell and isolate the surface proteins.

Three separate antigens were characterized by immunological and biochemical techniques and their morphology established by electron microscopy. One of the antigens was identified as being the 80 Å protein unit forming the faces and edges of the icosahedral shell. A second antigen was located at the 12 apices of the structure and linked with the third spike-like antigenic component. From the electron micrographs, coupled with biochemical and immunological data made available from many laboratories covering the period 1953–1976, it has been possible to build an accurate model of adenovirus at a resolution of about 25 Å. The progress made in determining the morphological features of human adenovirus is summarized in figure 4, plate III.

An important recent application of transmission electron microscopy in biology has been reported by Unwin and Henderson, who used high-resolution techniques to deduce the three-dimensional structure of the purple membrane. The purple membrane is a specialized part of the cell membrane of the rod-shaped bacterium, *Halobacterium halobium*, which functions as a light-driven hydrogen ion pump involved in photosynthesis. It contains one species of protein molecule (bacteriorhodopsin) of molecular weight 26 000, which makes up 75% of the total mass, and lipid which makes up the remaining 25%. Retinal, bound stoichiometrically to the protein, is responsible for the purple colour. These components together form a regular two-dimensional array based on a hexagonal lattice.

Like all biological specimens, the membrane is very susceptible to damage by electrons. This means that the resolution of an image of an isolated molecule exposed to an electron dose low enough to reduce damage to a minimum, is limited by electron noise. This limitation was overcome by combining the information from a large number of unit cells in the images of regular arrays so that genuine molecular detail was reinforced and the random statistical fluctuations arising from the electron noise were averaged out.

By collecting data from many such images and from electron diffraction patterns of the membranes, tilted through a series of angles with respect to the incident electron beam, Unwin and Henderson were able to obtain a three-dimensional map of the membrane to a resolution of 7 Å. Figure 5, plate IV, shows a model of the protein molecule based on the detail in this map.

3.2. *High Voltage Electron Microscopy*

The high voltage electron microscope (HVEM) enables thicker specimens to be examined than is possible at conventional (100 kV) voltages. This has important implications in both biological and materials applications—more information is available from a single picture and that information is more reliably characteristic of the bulk material since the effects of the specimen surfaces are decreased. This is particularly important in dynamic experiments, e.g. observation of the straining of a metal specimen. Alternatively the extra penetrating power may be used in surrounding a solid specimen with a gaseous or even liquid atmosphere and observing its behaviour in that environment.

One area in which the HVEM has been exceptionally useful, principally because of its greater penetration, is that of mineralogy, where large regions of samples must be examined to obtain meaningful results. An example of such an investigation is the study of Schiller colours in feldspars. It was suggested in 1923 that the cause of the Schiller colours (their characteristic iridescent colours in reflected light) in feldspars is the reflection of light by 'internal surfaces', the process being similar to the way that colours are produced on reflection of light from a thin film of oil on water. The HVEM was able to confirm this interpretation by showing that the reflecting surfaces are regularly spaced lamellar or prism-shaped particles of one feldspar component in a matrix of another (see figure 6, plate IV). The Schiller colour is related to the periodicity of the two components.

Another of the most important applications of the HVEM has been the study of atomic displacement damage. Electrons of sufficiently high energy can displace atoms from their sites in crystals by Rutherford collisions. The 1 MV electron microscope can thus produce atomic displacements in all elements up to an atomic weight of about 160, and the development of the radiation damage structure can be imaged directly as it occurs. The damage produced by fast incident electrons is similar to that produced by fast incident neutrons, hence HVEM studies produce information of relevance to practical nuclear reactor problems, in particular to the problems of the formation of voids in reactor materials. Far higher electron fluxes are obtainable in an HVEM than the flux of fast neutrons from a reactor and hence the radiation damage experiments using electrons yield information much more quickly than experiments using neutrons. Experiments can be performed in one day on the HVEM at an electron dose equivalent to neutron irradiation for several years in a fast reactor. An example of such a simulation is shown in figure 7, plate V.

Apart from the void problem there are various other aspects of radiation damage in which the HVEM plays an important role, for example the determination of the threshold energy for displacement damage in different materials and the migration energy for interstitial atoms.

3.3. Scanning Electron Microscopy

In the twelve years since the scanning electron microscope (SEM) first became commercially available in 1965, its performance has been greatly improved, its operation made easier, and its fields of application extended so that it is now used as a routine instrument for the examination of bulk specimens in all the fields of materials science and biology. With the development of brighter electron sources, operation at higher accelerating voltages, and improved lenses, stages and specimen holders, micrograph resolutions of ~ 50 Å are readily obtained.

In addition to contrast arising from changes in topography (see figure 8, plate VI, for examples) and atomic number, contrast characteristic of many other physical properties of materials can now be obtained. For example, magnetic contrast enables magnetic domains to be observed, while crystallographic contrast gives 'electron channelling patterns' and 'channelling micrographs' so that crystallographic orientations can be determined, grain and sub-grain boundaries observed, and information concerning local variations in dislocation densities obtained. Voltage contrast reveals variations in surface voltage distributions (figure 9, plate VII), cathodoluminescence reveals variations in luminescent properties, and electron beam induced conductivity (EBIC) reveals contrast variations in electrical properties within the specimens, these last three contrast mechanisms being of special importance for the investigation of semiconductor materials and devices. Images corresponding to all of these contrast mechanisms from bulk specimens can be directly displayed, several images can be observed simultaneously, and data can be quantitatively recorded and interpreted. Most of these contrast methods originated or were developed in the UK.

Dynamic experiments directly performed in the SEM in which the image is viewed and recorded while the specimen is progressively heated, cooled, deformed, fractured, subjected to changing voltages, electric or magnetic fields, melted, solidified, sintered, the surfaces mechanically worn, etc., are increasingly being used, the large specimen chambers allowing massive stages capable of complex functions to be constructed and operated. The use of beam-blanking procedures in which the electron probe is repeatedly switched on and off enables stroboscopic images of rapid periodically changing events, either electrical or mechanical, to be 'frozen' and directly observed. This technique was successfully applied to micro-electronic circuits (7 MHz) and microwave Gunn oscillators (~9 GHz). Automatic beam-tilting procedures allow stereoscopic images of topographical features to be directly and continuously viewed.

A basic principle of the SEM, namely, that the collected signal varies as a function of time, enables electrical signal processing of each picture point in the image to be automatically performed. Improved images can then be obtained with increased contrast, reduced noise, etc., while the use of an on-line mini-computer provides automatic instrument control, recording of data, and interpretation of results. The SEM is increasingly being modified so that it can also be used for analytical work. The most common addition is the incorporation of energy or wavelength dispersive characteristic X-ray detectors for 'bulk' chemical element analysis (see next section), while the detection of Auger electrons for 'surface' chemical element analysis is also being used.

A specialized and rapidly growing field of application of the SEM is for micro-fabrication in the semiconductor electronics industry. The fine electron probe is automatically controlled so that it exposes electron-resist surface films on silicon micro-electronic chips to produce masks with edge resolutions superior to those produced by light optical methods. Pioneering work in this field was performed in the UK. An example of 'microwriting' using this technique is shown in figure 10, plate VII.

Although the major applications of the SEM have been with bulk specimens, increasing use is now being made of scanning transmission electron microscopy (STEM) with thin foil specimens. With the aid of a field emission electron source, micrograph resolutions of 3 Å are now obtained. The most useful application of this technique is for micro-analytical work. The large electron currents that can be produced in fine probes enable X-ray chemical element analysis, and electron energy loss plasmon and K-shell analysis, to be obtained from specimen areas 100 Å across, and transmission electron diffraction patterns from areas 30 Å across.

3.4. X-ray Microanalysis

Scanning electron probe microanalysis originated in the mid-1950's (see §§2.2 and 2.3) and its potential was quickly realised following its successful application to several important metallurgical problems. Modern instruments are more sophisticated in design, with an associated enhanced performance compared with their predecessors, but the underlying principles remain the same. The aim is to obtain detailed information about the chemical composition of very small regions in the surface (usually polished and lightly etched) of a thick specimen, typically a few millimetres in diameter and thickness, particularly information on the type and amount of elements present and their distributions. This is achieved by directing a finely focused, high-energy (5–40 keV) electron beam at the specimen where it excites X-rays from a region approximately 1 micrometre in diameter at the specimen surface. Spectral analysis of the characteristic X-rays generated with the electron probe stationary at a selected spot enables the complete chemical composition of a micro-region to be obtained, while scanning the probe over an area reveals the distribution of elements in the surface layer in the form of a distribution map. Quantitative analysis is performed by comparing the X-ray intensities from the specimen with those from standards of known composition. Quantitative data are obtained after certain 'matrix corrections' have been applied. The physical basis for these corrections has received much attention in the last two decades with the result that the accuracy is now usually better than $\pm$ 2% (relative), which is adequate for most applications.

One of the very early metallurgical applications may be used to exemplify the technique. The root cause of surface cracking of mild steel tubes subjected to certain hot-working conditions was identified in quantitative terms involving the enrichment of certain residual elements in the sub-scale region beneath the oxidized surface (figure 11, plate VIII). The micrograph labelled 'electron' was the image obtained by employing the electrons backscattered from the specimens. The five other pictures are 'X-ray distribution images'—the light areas were regions rich in the elements cited. It will be noted that copper, tin, nickel, antimony and arsenic were all concentrated in the subscale.

A myriad of metallurgical applications has followed in the wake of this pioneering work and the technique is now in worldwide use for routine identification of inclusions, second phases and precipitates, impurities, and the distribution of segregated alloying elements. The technique has been particularly useful in the study of minerals, as well as fruitful in certain fields of biology.

An offspring of the scanning electron probe microanalyser was the second generation instrument known as EMMA (see § 2.3). An important feature is that thin specimens are used, in contrast to the much thicker 'bulk' specimens studied by conventional electron probe microanalysis. The greatly reduced electron scattering within thin specimens allows X-ray microanalysis to be performed with a resolution of 0.1–0.2 μm (for comparison, the unavoidable sideways scattering of electrons as they enter a thick sample limits the resolution to about 1 μm).

EMMA has found abundant application, both in metallurgy and biology, in studies of thin sections and extracted particulate matter. A good illustration of the value of the improved resolution of the compositional analysis was provided by the studies of heterogeneous nuclei for graphite nodules in chill-cast iron, and a specific example is shown in figure 12, plate IX. The nuclei were hexagonal-shaped plates, about 1 μm in diameter and 0.1–0.3 μm thick. A proportion of the potential nucleating particles were rendered inactive (i.e. 'poisoned' as active nuclei) because of a thin layer of surface adsorbed sulphur about 15 Å thick, which could be detected with the crystal spectrometers. In other respects the active nuclei were identical with the inactive particles, figure 12, for which dark field microscopy indicated a duplex structure with a central particle ~ 500 Å in diameter. The energy-dispersive X-ray detector showed Mg, Al and Si in the outer region, with these elements plus Ca and S at the centre. Additional compositional and diffraction information led to the conclusion that the central particle was a mixed sulphide (Ca, Mg)S with a specific crystallographic relationship with the surrounding single crystal of spinel-type (Mg, Al, Si) oxide whose structure was suitable for overgrowth of graphite, i.e. in the absence of sulphur 'poisoning' the particle was an ideal heterogeneous nucleus for graphite.

X-ray microanalysis has also led to important advances in biology. For example, before the advent of the technique it was not possible to measure accurately the local concentrations of diffusible elements in cellular components and narrow extracellular spaces in biological tissues. The method of electron microprobe X-ray analysis of tissue sections about 1 μm thick in a deep frozen and hydrated state, as developed by Hall and his colleagues, permits fully quantitative measurements of various elements in morphologically recognizable components with a resolution of 2000 Å or better. The type of information obtained is illustrated in figure 13 by some results from the salivary glands of *Calliphora*, stimulated to secrete fluid. Even more detailed information has been obtained from several other tissues.

A good measure of the flexibility and impact of any analytical technique is the extent to which it can assist with topical problems. Pollution from water and air-borne particles is currently receiving much attention. EMMA is playing its part in this work, for example as a technique capable of distinguishing between different types of very small asbestos fibres.

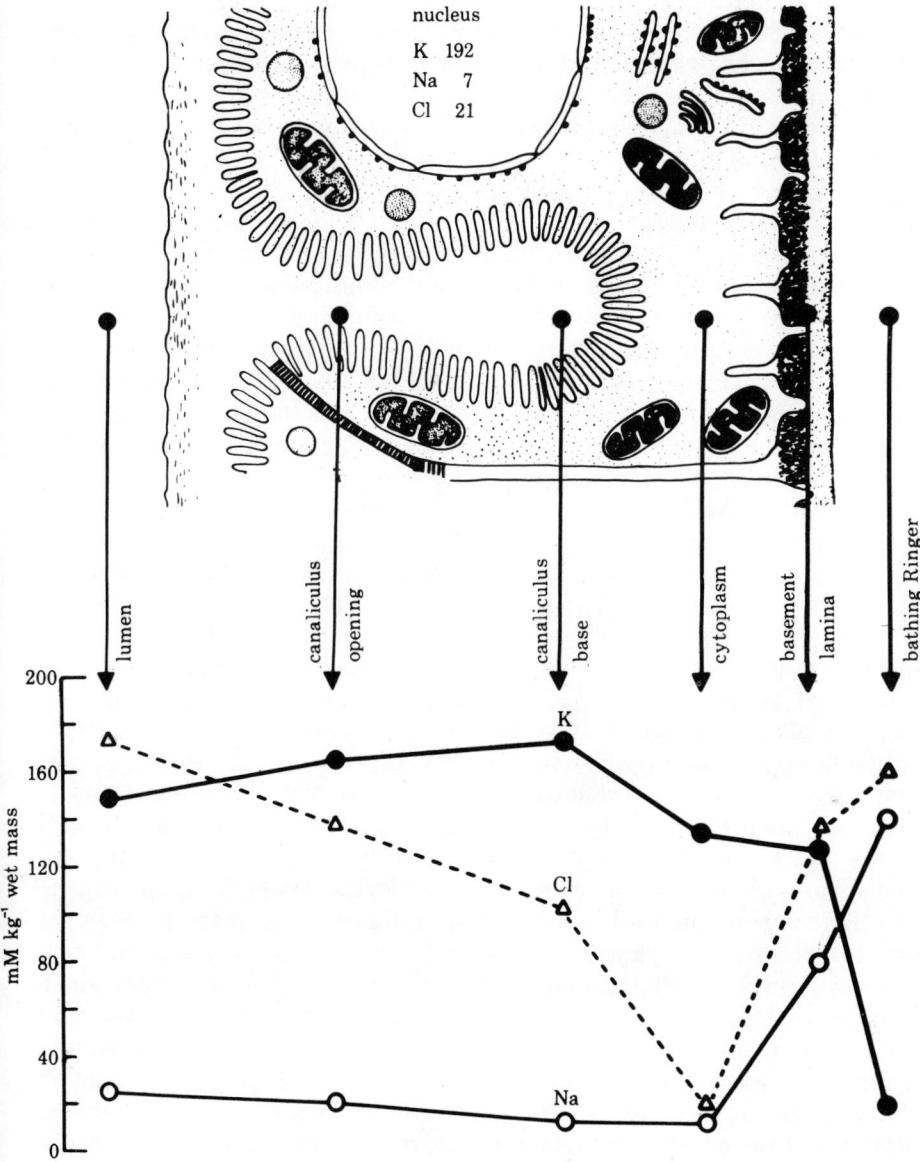

FIGURE 13.   Summary of results of X-ray microanalysis studies of frozen hydrated sections of the salivary glands of *Calliphora*, stimulated with 10^{-8} M 5-HT to produce isotonic saliva rich in KCl. The most interesting feature is the very high concentration of KCl in the caniculus which is thought to be the site of solute-water coupling by local osmosis (B.L. Gupta, T.A. Hall and R.B. Moreton).

4. Some Future Developments

The present trend of instrumental development is towards the production of electron optical instruments of extreme versatility, combining in one instrument the functions of conventional high resolution transmission electron microscopy, scanning electron microscopy, scanning transmission electron microscopy, X-ray microanalysis, and electron energy loss spectrometry. The invention of electron sources of high brightness, such as the lanthanum hexaboride and field emission sources, has greatly aided this development.

Progress towards higher resolution is dependent on improvements in electron lens design in order to minimize aberrations. At present electron lenses are almost completely uncorrected. Spherical and chromatic aberrations could be corrected in principle (a) with axially unsymmetrical elements such as quadrupoles or octopoles, or (b) by use of pulsed lenses at GHz frequencies. Work at Cambridge has shown that method (a) is more practicable than method (b) for reducing spherical aberration. At Darmstadt an achromatic lens system is now in operation. Other aberrations would then stand in the way of achieving the wavelength-limited resolving power (< 0.037 Å at 100 kV), but 0.5 Å is a reasonable target. Such an image resolution would be of great value in solid state research, but in molecular biology radiation damage prevents the use of a sufficiently high magnification. Present lenses, however, in a highly stabilized high voltage electron microscope, can in principle provide a resolving power approaching 1 Å, which would image the structure of some radiation resistant molecules if low exposure techniques of image recording were employed.

Instead of attempting to correct lens aberrations directly, it is possible to process an aberrant image so as to remove their confusing effect *a posteriori*. Because this procedure may give rise to artifacts in the reconstituted image, it is being thoroughly investigated at the Cavendish Laboratory on both test models and actual micrographs. High resolution image detail varies rapidly with defocus. There is no 'right' focus position: the image is faithful for certain levels of detail only, and misleading for others; which levels are which change with the focus. By computer processing we can now build up a true image by retaining from each of a series of pictures only those levels of detail imaged faithfully at that focus position. With experience, this approach should yield images with a resolution as good as instrumental instabilities and residual aberrations will allow, and certainly of the order of 1 Å. To make proper use of such a possibility the specimen would have to be cooled down to liquid helium temperatures, in order to minimize the effect of thermal agitation. Whether the conventional electron microscope or its scanning transmission counterpart will first attain such a performance must remain an open question.

Even when image processing or lens correction has given a reliable 'true' image of the specimen, there remains the important problem of how to interpret the image in terms of structure. For crystalline specimens with periodic structure this problem does not seem as severe as it is for amorphous specimens, and examples have been given recently where the crystallographic structure of

FIGURE 3. Lattice fringes in platinum phthalocyanine, with a spacing of 12 Å (J.W. Menter).

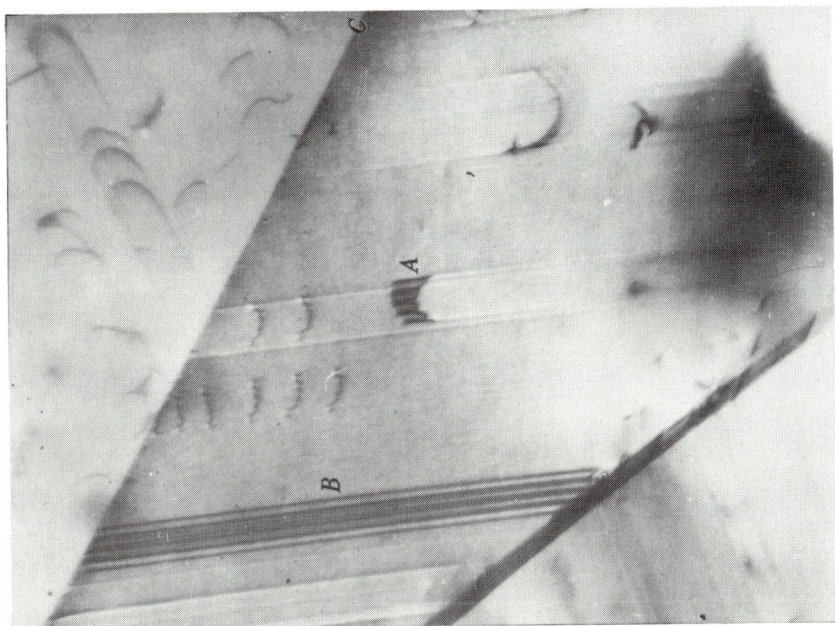

FIGURE 1. Transmission electron micrograph of a thin foil of stainless steel showing: A—dissociated dislocation; B—wide stacking fault; C—twin boundary (P.B. Hirsch, R.W. Horne, M.J. Whelan and W. Bollmann) Magnification 45 000.

PLATE II (JENKINS & WHELAN)

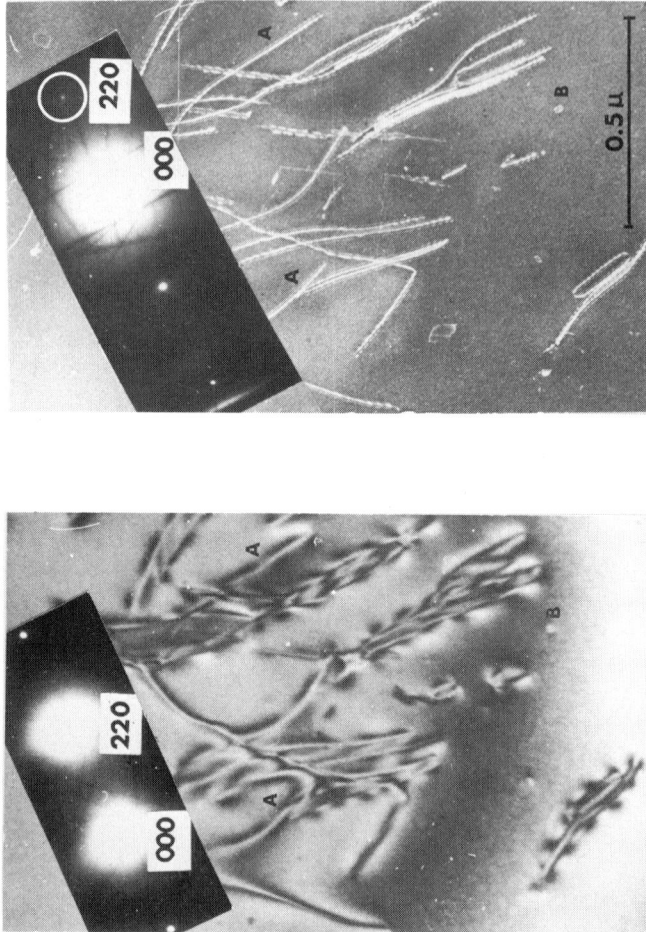

FIGURE 2. (a) Strong-beam 220 dark field image of silicon specimen containing dislocations; (b), correspondingly weak-beam 220 dark field image (D.J.H. Cockayne and I.L.F. Ray). Magnification 56 000.

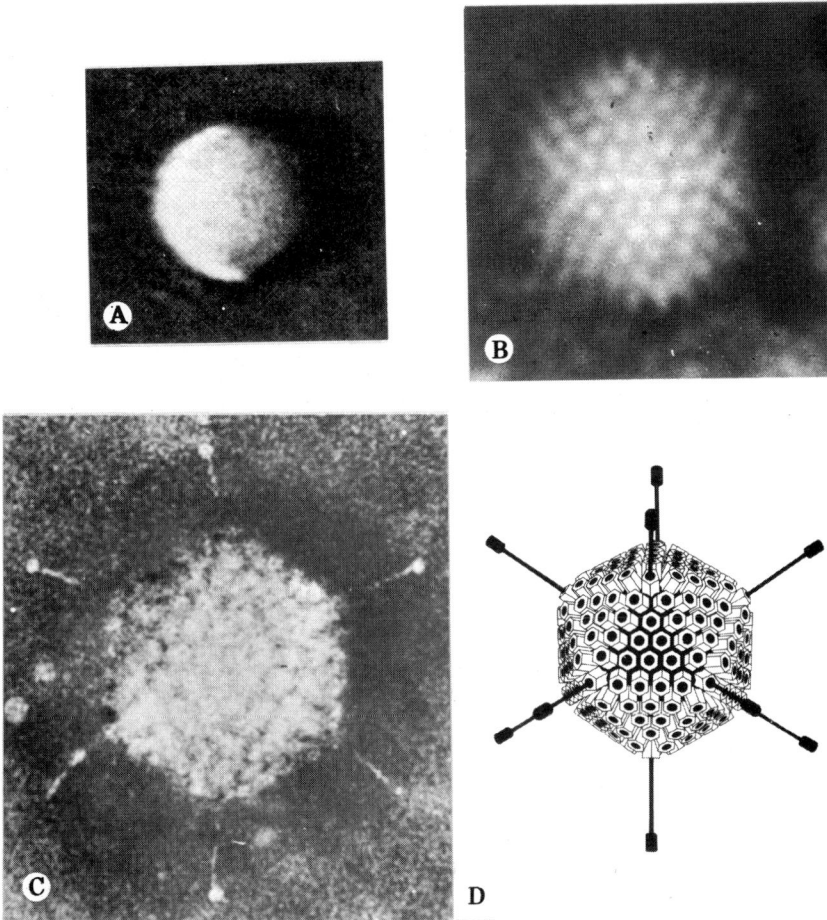

FIGURE 4.   The electron micrographs show the progress made in determining the morphological features of human adenovirus.
A. Shadowed preparation of an isolated particle of adenovirus. The information is limited to the approximate shape and size of the virus (R.C. Valentine, 1957). Magnification 200 000.
B. Negatively stained preparation of adenovirus showing the distribution of 252 protein units to form an icosahedral shell (R.W. Horne *et al.*, 1959). Magnification 700 000.
C. With improved preparative methods the 12 spike-like projections are shown at the apices of the protein icosahedral shell (R.C. Valentine and H.G. Pereira, 1965). Magnification 400 000.
D. Model of human adenovirus showing the distribution of the surface protein components (R.W. Horne, I. Pasquali-Ronchetti and J.M. Hobart, 1976).

PLATE IV (JENKINS & WHELAN)

FIGURE 6. The periodic structure in a moonstone from Burma which gave rise to a white Schiller colour (G.W. Lorimer and P.E. Champness). Magnification 12 000.

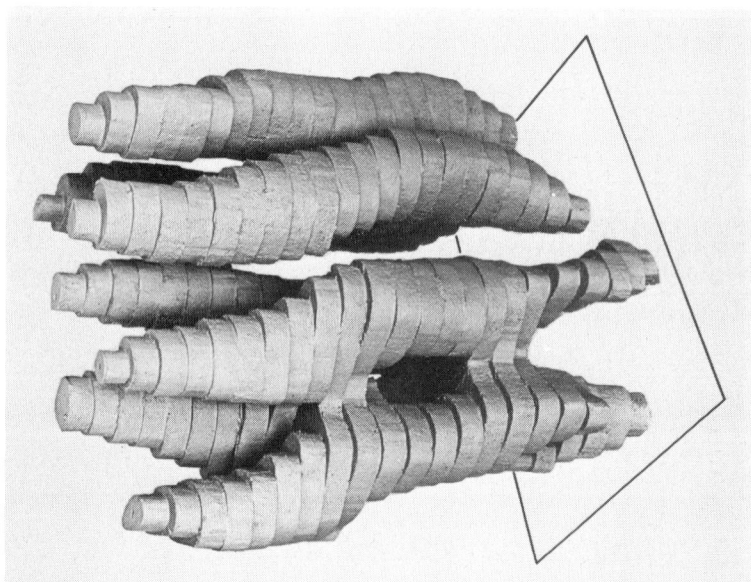

FIGURE 5. Model of a single protein molecule in the purple membrane viewed roughly parallel to the plane of the membrane. The seven rod-shaped features correspond to α-helices, each 35–40 Å in length (R. Henderson and P.N.T. Unwin).

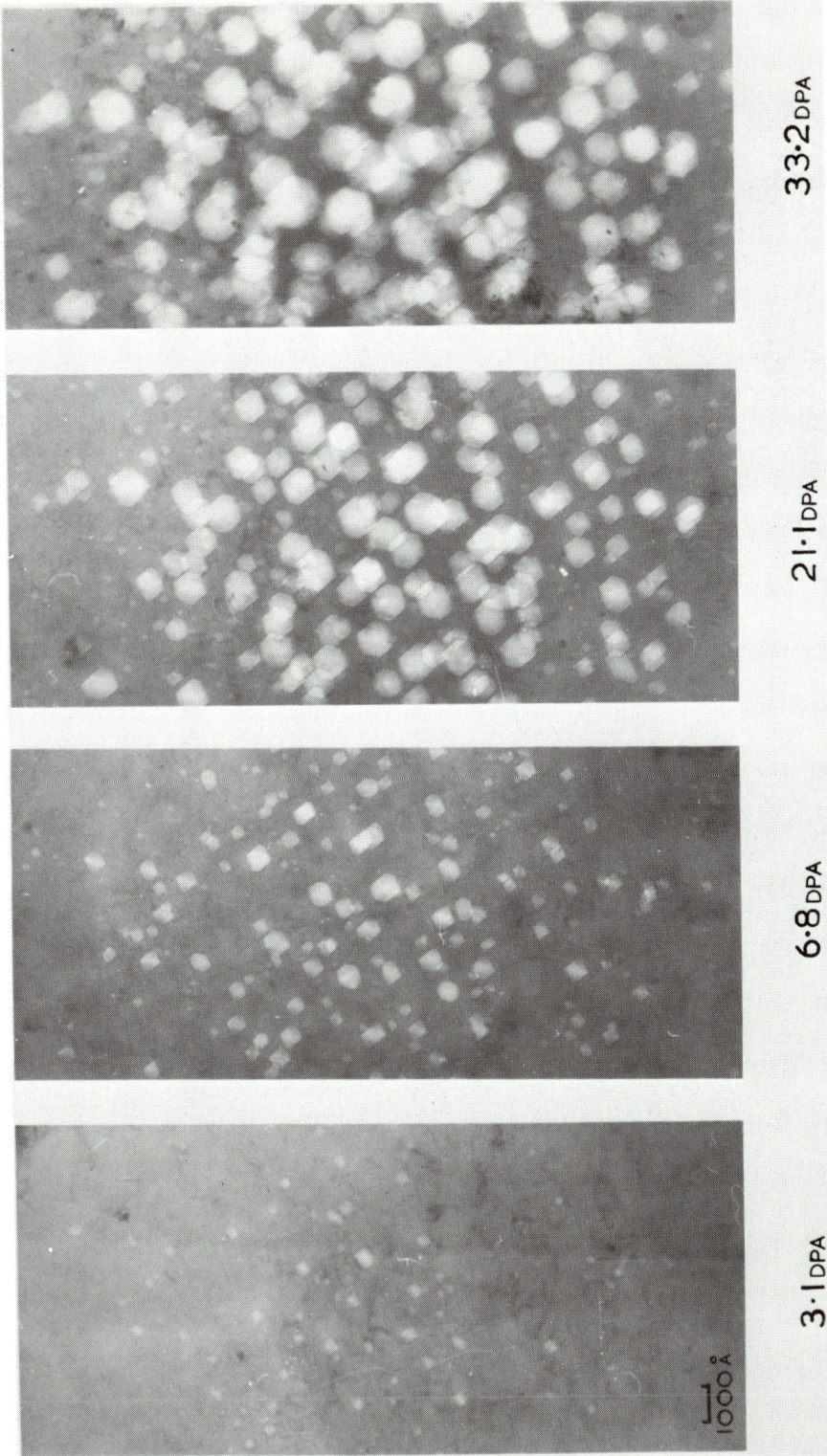

FIGURE 7. Void formation and growth in Type 316 austenitic stainless steel, electron irradiated at 600°C. The figures of displacements per atom (DPA) refer to the number of times each atom has been displaced, on average, from its normal lattice site. 1 MeV electrons. AEI HVEM (M.J. Makin). Magnification 50 000.

PLATE VI (JENKINS & WHELAN)

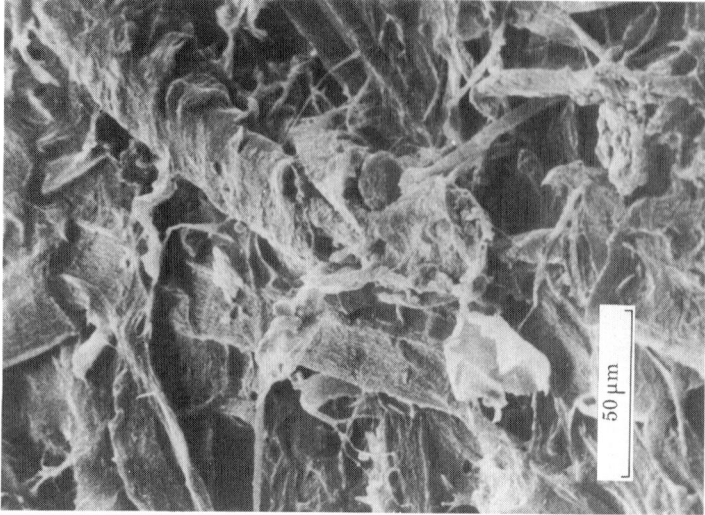

(b) Surface of newsprint taken in the third Cambridge scanning electron microscope (K.C.A. Smith).

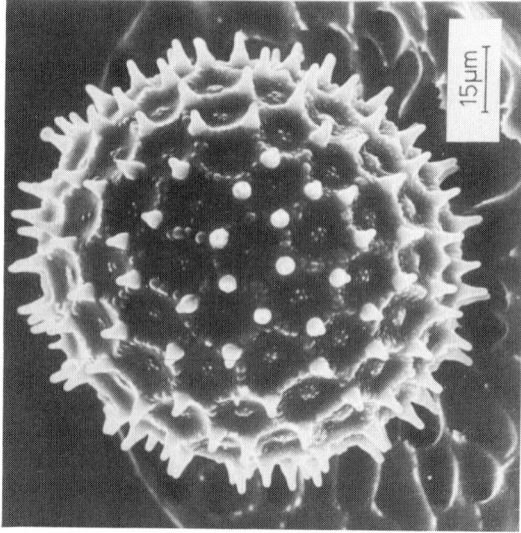

(a) A pollen grain of the Morning Glory, *Ipomoea purpierea* (P. Echlin).

FIGURE 8. Topographical contrast in the scanning electron microscope

FIGURE 9. Voltage contrast in the scanning electron microscope—direct observation of voltage differences on the surface of MOS Ladder Circuit. 5 volts dc on top electrode (G.S. Plows and W.C. Nixon).

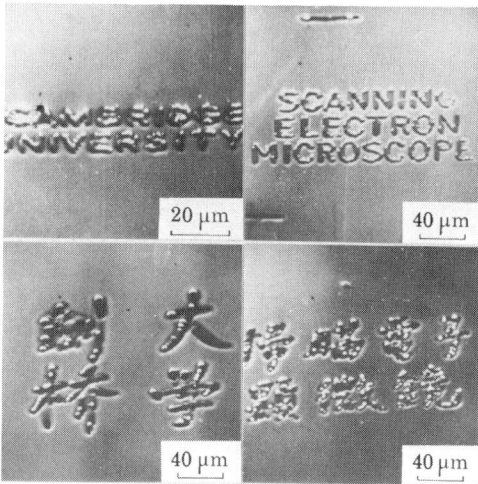

FIGURE 10. Electron beam lithography: microwriting in silicon (T.H.P. Chang and W.C. Nixon).

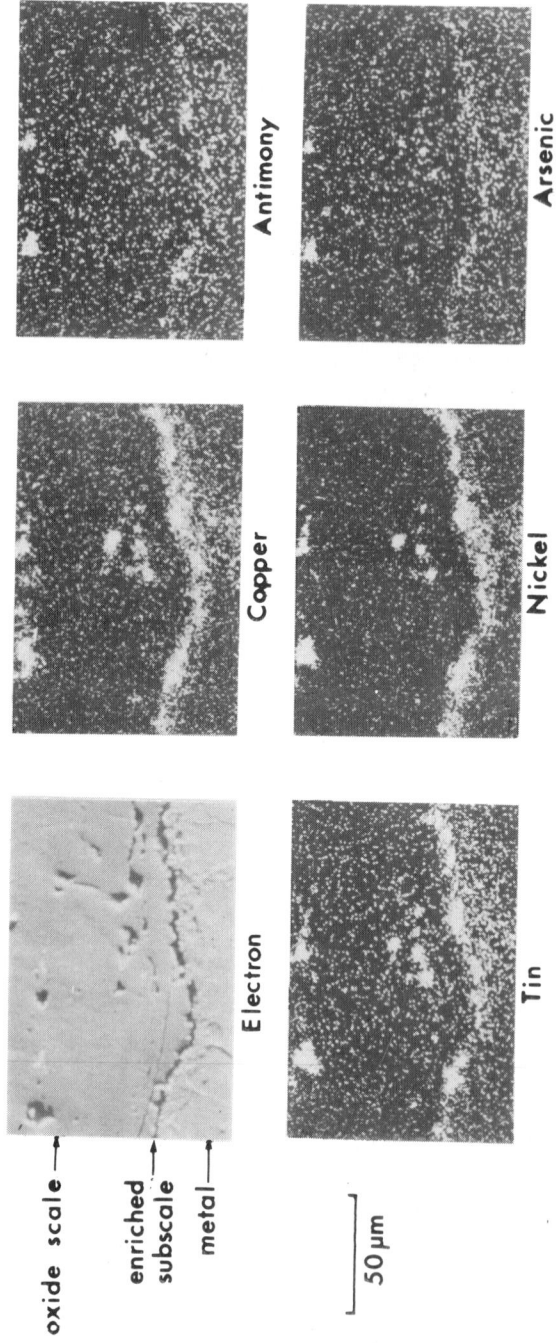

FIGURE 11. Scanning electron-probe micrographs of normal section through oxidized mild-steel surface showing sub-scale enrichment of residual elements (D.A. Melford).

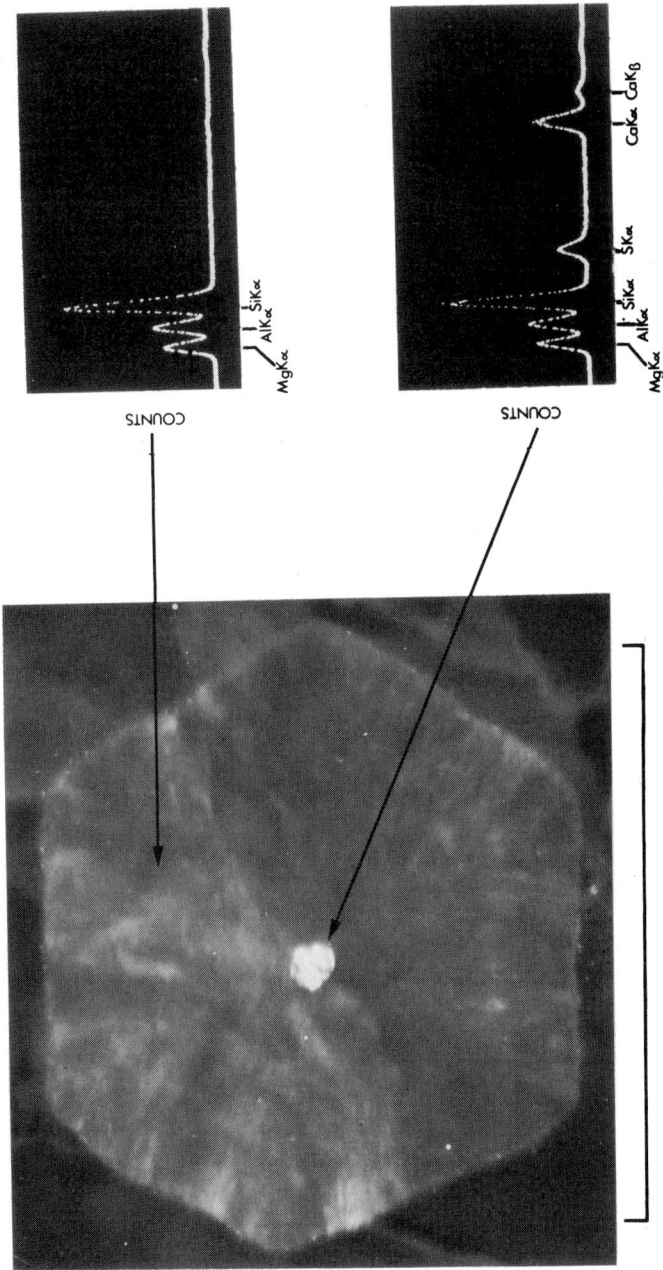

FIGURE 12. Microanalysis with EMMA, employing a 0.1 μm/100 kV electron probe, of an extracted duplex particle from chill-cast iron (M.H. Jacobs).

inorganic compounds has been unravelled speedily by direct electron microscope examination. Undoubtedly there will be more applications of this method by chemists and crystallographers in the future.

CHEMISTRY IN MICROTIME

SIR GEORGE PORTER, F.R.S. AND M.A. WEST
The Royal Institution, London

INTRODUCTION

One of the principal activities of man as scientist and technologist has been the extension of the very limited senses with which he is endowed so as to enable him to observe phenomena with dimensions very different from those he can normally experience. In the realm of the very small, microscopes and microbalances have permitted him to observe things which have smaller extension or mass than he can see or feel. In the dimension of time, without the aid of special techniques, he is limited in his perception to times between one twentieth of a second (the response time of the eye) and about 2×10^9 seconds (his lifetime). Yet most of the fundamental processes and events, particularly those in the molecular world which we call chemistry, occur in milliseconds or less and it is therefore natural that a chemist should seek methods for the study of events in microtime.

The extension of the time range of chemical experiments has been one of the most important post-war developments in physical chemistry. Twenty-five years ago, a chemist who heard the words microsecond and nanosecond was probably in the wrong laboratory. The familiar ionic reactions of inorganic chemistry produced colour changes or precipitates 'instantaneously' and the reactions were described as 'immeasurably fast'. Although there was good indirect evidence for all kinds of short-lived, unstable intermediates in chemical reactions, few of these intermediates had been directly observed and there were often as many mechanisms proposed for a particular reaction as there were workers in the field. Today, however, although the chemist retains some of his ancient rights to speculate on how a reaction goes and on what intermediates are involved, he is considerably restrained by the knowledge that those transient intermediates can probably be observed and measured during their short existence as certainly as can the stable products of the reaction.

The rates and durations of chemical change vary over a wide range as is shown in figure 1. Reactions whose progress can be followed by conventional methods in the laboratory are limited to the middle part of the range. Very much slower reactions, such as rusting, corrosion and decomposition, are, of course, often quite important and nearly every substance in the chemical laboratory is

155

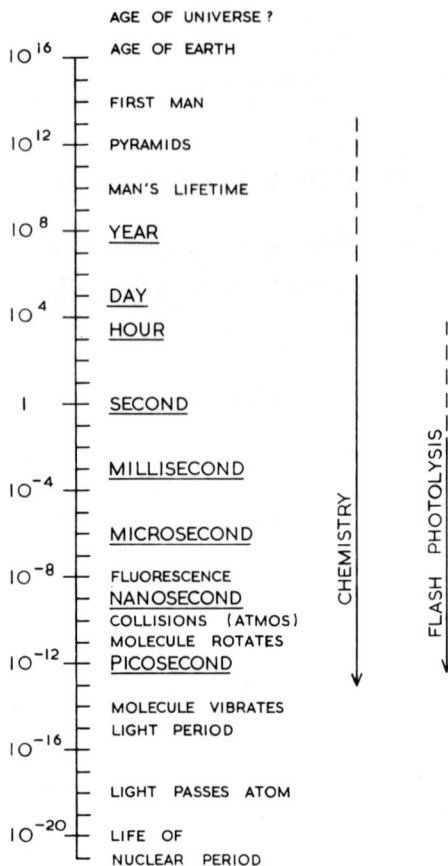

FIGURE 1.   A time scale, in seconds, showing the
range of chemistry and of flash photolysis.

thermodynamically unstable and is decomposing at a slow but definite rate. At
the other end of the scale where chemical changes take place in less than one
second, there is a much greater variety of important reactions. Indeed the study
of chemical kinetics is almost synonymous with the study of fast reactions
because the simplest and most fundamental reactions are usually very fast and
most slow reactions proceed by a series of steps many of which are also fast. The
observational gap which has to be filled extends from one second to about 10^{-14}
second and over this time scale there are two crucial experimental problems
which must be overcome in order to investigate any fast chemical reaction. First,
the reaction must be initiated in a time that is short in comparison with the overall
reaction time, and in a manner that ensures homogeneity of reaction throughout
the volume of reaction mixture that is studied kinetically. Second, a significant
parameter of the reacting system must be recorded, after initiation, in such a
manner that useful information regarding the mechanism and the kinetics of the

reaction can be derived therefrom.   There are several general methods of bringing about a chemical reaction (e.g. by mixing the reactants rapidly together, by heating the reaction mixture, or by subjecting the reaction mixture to radiation or an electrical discharge). For very fast reactions, some of the methods have fundamental limitations. The mixing of two reactants is a powerful method (especially the variation known as stopped-flow) and, with most liquids, can be brought about in a time of about 1 ms. Obviously rather longer times are required for viscous liquids and gases and the method is not appropriate for solids.   Initiation of chemical change by heating a reaction mixture in a conventional way is a very slow process. In a gas, at atmospheric pressure and in a vessel 1 or 2 cm inside diameter, the time taken for a temperature change at the wall of the vessel to be equilibrated through the gas is of the order of seconds; naturally the time is much longer in liquids. The temperature-jump relaxation method makes it possible to raise the temperature of a liquid in a time of the order of microseconds; other relaxation methods (of comparable time resolution) utilize pulses of pressure or electric field.

The initiation of reaction by irradiation is less common, for preparative chemistry, than heating or mixing but it has proved to be the most powerful method for the study of chemical mechanisms and of transient intermediates of chemical change. It is therefore fortunate that photochemical initiation of reaction (by a technique known as flash photolysis) is the most convenient method of bringing about a large extent of reaction in a short interval of time in a homogeneous system.

FLASH PHOTOLYSIS

Flash photolysis is a method whereby a dramatic non-equilibrium situation can be created in a reaction system in a short interval of time. A material under investigation is subjected to a short, intense light flash of visible or ultraviolet light. The intensity of the flash must be sufficient to produce a change in chemical composition that is measurable but of short duration compared with that of the ensuing reaction. This technique has proved to be a general means of preparing unstable chemical intermediates in concentrations higher than can be obtained by other methods, which rarely provide non-equilibrium concentrations of species such as free radicals high enough for direct observation. Along with the sister technique of pulse radiolysis, it is almost the only method available for the preparation and study of high concentrations of electronically excited molecules. It is applicable to gases, liquids and solids and to the whole available temperature range. The non-equilibrium situation created by the flash can be made homogeneous throughout the reaction vessel of any required shape and dimensions. In conjunction with physical methods of observation, the concentrations of intermediates can be directly measured as a function of time, and their chemical and physical properties determined.

Like the photochemical method itself, flash photolysis is of very general applicability and, with an appropriate choice of reacting system, most types of

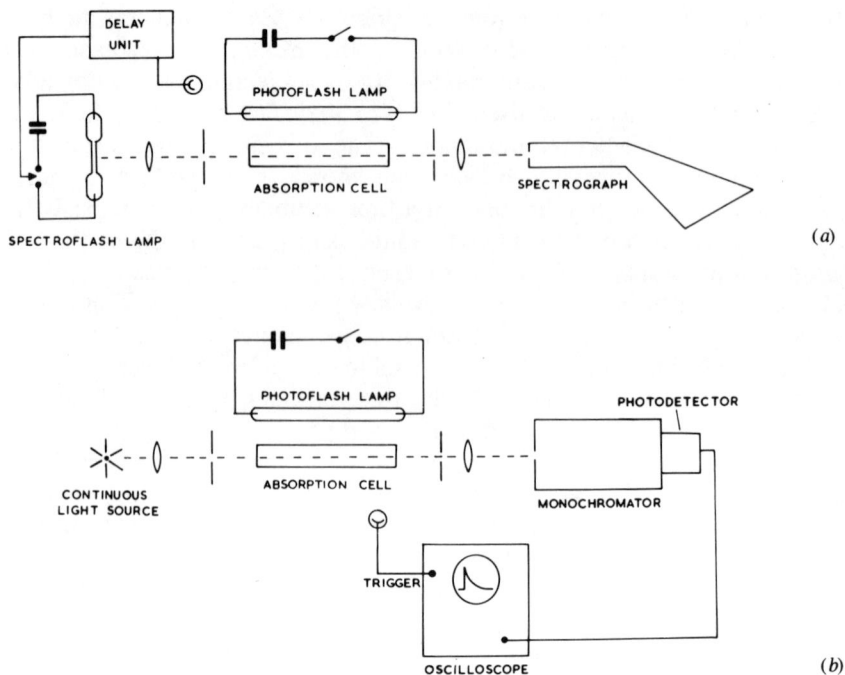

Figure 2. Schematic diagrams illustrating principles of (a) flash spectroscopy and (b) kinetic spectrophotometry.

chemical change can be initiated and most classes of chemical intermediates prepared. On the other hand, photochemical techniques are rather specific and, by a suitable choice of wavelength, the primary process can often be very clearly defined.

Information on chemical intermediates produced by flash photolysis is usually obtained from absorption spectroscopy in the ultraviolet and visible regions. Transient absorption spectra can be investigated in two ways and a single experiment can serve to record either the whole range of wavelengths at a single time or alternatively a single wavelength over all times. The experimental arrangements for these techniques, which are known as flash spectroscopy and kinetic spectrophotometry, respectively, are shown in figure 2. Each technique has its advantages and disadvantages and the principal characteristics are shown in table 1. For exploratory work on a chemical system in which the spectra of the transients are unknown or uncertain, it is advisable to record the spectra photographically over the whole spectral range of interest. For gas phase studies, where simple intermediates having fine structure are likely to appear, photographic recording on a spectrograph is almost essential. Kinetic spectrophotometry is used to obtain information on the decay processes since the whole kinetic record of concentration versus time is obtained in a single experiment.

TABLE 1

Principal Characteristics of Flash Spectroscopy and
Kinetic Spectrophotometry

Characteristic	Flash spectroscopy	Kinetic spectrophotometry
Sensitivity (% absorption change detectable)	5–95	0.1–100
Spectral range (nm)	100–800	100–2500
Spectral resolution (nm)	typically 0.05 nm	typically 2–5 nm
Production of transient spectrum	easy usually one flash	difficult for high resolution spectra requires many flashes
Kinetic analysis	difficult for any decays but first order requires many flashes	easy usually requires only one flash

In the original apparatus designed for gas phase studies, a flash energy of 10 000 J was discharged in a few milliseconds and synchronization between the photolysis and spectroscopic flashes was effected by a rotating mechanical shutter. The reaction vessel was 1 metre long and the high efficiency of light production resulted in almost complete destruction of substances like acetone and NO_2 in the gas phase. Most subsequent work has utilized smaller energies of the order of several hundred joules with shorter pathlength cells.

Since all compounds absorb light and almost any type of reaction can be initiated by light, it is apparent why the technique has such great potential for studying fast chemical reactions. The following examples of chemical systems studied by conventional flash photolysis are given before a discussion of recent applications of lasers as excitation sources. The laser has transformed the time resolution of the technique to a point not far from the uncertainty limits of chemical systems.

FREE-RADICAL SPECTROSCOPY

At the time that the flash photolysis method was introduced, no absorption spectrum of a polyatomic free radical was known and the earliest flash spectroscopic record shows what occurs in a mixture of chlorine and oxygen when it is irradiated with light (figure 3, plate I). There is no overall photochemical reaction in this mixture of gases but immediately after flash irradiation a brand new spectrum appears which has been shown to be the absorption spectrum of the diatomic radical ClO. When a new transient species is detected in this way two kinds of information become available. First, analysis of the spectrum itself leads to structural and energetic data about the substance; secondly, studies of the time-resolved spectra provide a measure of its concentration as a function of time and therefore provides kinetic data about the physical and chemical changes which it undergoes. Chlorine monoxide was the precursor to the discovery of many new radical spectra including nearly all the ones of principal chemical and physical interest. For example, Herzberg, working in the first place in the vacuum ultraviolet, was able to observe the methylene and methyl radicals, to say something of their structure and to resolve, to a great extent, the problem of the triplet and singlet states of methylene.

The other particularly fruitful field under this heading has been that of the aromatic free radicals. It seems that, given the right conditions, almost every aromatic molecule can be made to yield a free radical spectrum by flash photolysis. The spectra are, of course, far more complex than those from small molecules but many of them are nevertheless quite sharp and show much fine structure in the vapour phase. Phenyl radicals were observed following flash photolysis of benzene and halogenated benzenes and, surprisingly, cyclopentadienyl radicals (C_5H_5) identified from the flash photolysis of aniline, phenol, nitrobenzene and many other substituted benzenes. The latter discovery points to the intriguing photochemical problem of the primary process in the excited states by which these transformations take place. In the case of C_5H_5 it seems probable that a biphotonic process involving radicals such as phenoxyl, benzyl and similar radicals may be operative.

EXCITED MOLECULES

In addition to the spectra of free radicals which, although chemically unstable, are physically stable, flash photolysis has recorded the spectra and kinetics of a variety of physically unstable molecules. In 1944, Lewis and Kasha showed that the phosphorescence of organic molecules which is observed in rigid media is the emission of light from the lowest excited state of these molecules and that this state is of triplet multiplicity. The influence of this discovery on chemistry and photochemistry was at first very slight. The reason for this was that the chemist does little work in rigid solutions at low temperatures and only under these conditions could the triplet states be observed. The absence of phosphorescence from gases and liquid solutions was undoubtedly due to the much shorter lifetime of the triplet state and short-lived absorption spectra of aromatic molecules in their triplet states were first observed by flash photolysis in 1952. It was these studies, perhaps more than any others, which brought flash photolysis into the chemical laboratory as a routine method of investigation (figure 4). Any discussion of mechanism in organic photochemistry immediately involves the triplet state and questions about this state are most directly answered by flash photolysis. It is now known that many of the important photochemical reactions in solution, such as those of the ketones and quinones, proceed almost exclusively via the triplet state. Its relatively long lifetime, compared with the time of a flash experiment, has made it possible to study the triplet state almost as readily as the ground state, and in many systems its physical properties and chemical reactions are now as well characterized as those of the ground state.

As examples of chemical processes in the triplet state we may mention proton

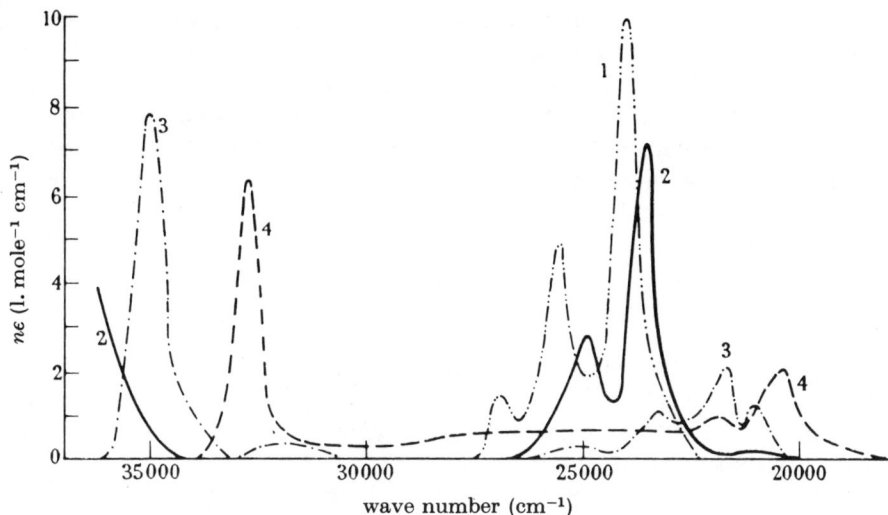

FIGURE 4. Triplet-triplet absorption spectra of the linear polyacenes. 1, naphthalene; 2, anthracene; 3, naphthacene; 4, pentacene.

transfer, and electron or hydrogen atom transfer. It is usually easy to arrange, by using buffered solutions, that protonic equilibrium is established during the lifetime of the triplet and, in this case, a titration can be carried out almost as readily as when one determines the pK of the ground state though now the 'indicator' is the molecule in the triplet state. Flash photolysis studies of electron and hydrogen atom transfer, particularly from solvent to triplet states of ketones, aldehydes and quinones revealed some fascinating properties of benzophenone derivatives. It was found that the lowest triplet state is one which always reacts and the electronic structure of the state is therefore the prime consideration. Depending on the substituents and solvent, this lowest triplet state may be $n–\pi^*$, with electrophilic oxygen and therefore reactive, or $\pi–\pi^*$, with considerable charge transfer character in the opposite sense to that of the $n–\pi$ state and therefore unreactive.

These studies illustrate how the excited electronic state must be treated as a new species, with its own structure. Since each molecule has only one state, but several excited states, it is clear that this field of investigation is, in a real sense, a bigger subject than the whole of conventional chemistry.

KINETIC INVESTIGATIONS

The first application of kinetic spectrophotometry following flash photolysis was to the study of the kinetics of iodine atom recombination. It was first necessary to show that the recombination of photolytically produced atoms was indeed a third-order reaction, as had been predicted theoretically for many years.

A number of interesting complications and problems arose with this study. Deviations from linearity were found which were caused by the unexpected high efficiency of the iodine molecules as third body. There were striking differences in efficiency between different third-body molecules and temperature coefficients were negative. All these observations were eventually explained in terms of a mechanism involving intermediate formation of a complex between the iodine atom and third body, M:

$$I + M \rightarrow IM$$
$$IM + I \rightarrow I_2 + M$$

In some cases (e.g. NO) the bonding in the complex is undoubtedly chemical but in others (e.g. He, O_2 or CH_3I as third body) complexes are suggested to be of the charge-transfer type and spectroscopic evidence is obtained from further flash photolysis studies of iodine atoms.

Nowadays, kinetic measurements following flash photolysis are extensively applied to large molecules in solution and there can be few reactions in organic photochemistry where such measurements do not assist in the elucidation of the mechanism and in providing new quantitative data on rate constants of the intermediate reactions involved. In principle, the investigator should observe the first excited state reached by absorption, lower states reached by radiationless

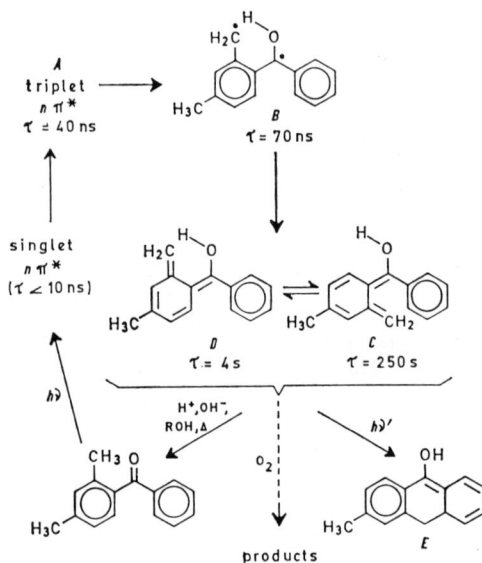

FIGURE 5. Intermediates and reaction pathways of photoreaction of 2,4-dimethyl benzophenone.

conversion, isomers, free radicals and all other intermediate stages in the photochemical transformation. This ideal state of affairs is rarely attained: some of the intermediates may have lifetimes shorter than the resolving times of any method yet available (a situation which is fortunately becoming rarer with laser techniques) and some of the intermediates may give weak absorptions or overlapping spectra. Nevertheless, in some fairly recent work, as many as five transient intermediates have been observed and followed kinetically in the transformation of a single substance (figure 5).

The cage effect often reduces the quantum yield of radical formation in solution but this is compensated for by the much higher concentration of higher molecular weight compounds which can be obtained in solution. In addition to radical production by dissociation, they may be formed by reaction of the parent molecule with solvent or other substrate, and a large variety of radicals of the semiquinone and ketyl type, produced by reaction of triplet carbonyl compounds, has been studied in this way. If a molecule does not absorb light in a convenient region, the excitation may be brought about by energy transfer from another molecule which has been electronically excited (e.g. excitation of vinyl compounds and cis–trans isomerization).

Examples of unstable substances which have been produced and observed by flash photolysis are shown in figure 6.

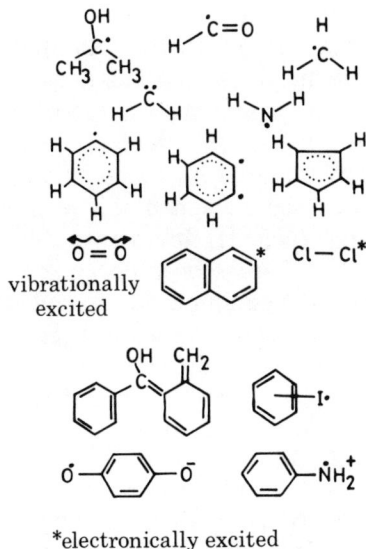

FIGURE 6. Unstable substances which have been observed and studied by flash photolysis. Most of the examples shown here normally have lifetimes below 1/1000 second.

Fluorescence Lifetime Measurements

The examples so far have all been taken from conventional flash photolysis studies employing high-energy flash lamps for excitation. Unfortunately, the time duration of a flash is related to its energy and it has proved very difficult to produce flashes of less than one microsecond duration with sufficient energy to produce a measurable absorption change in a normal flash photolysis experiment. However, repetitive flashes of nanosecond duration are extremely useful for measuring fluorescence decay profiles in the submicrosecond region.

Light is produced from a 1 mJ discharge between electrodes placed a few millimetres apart and serves two functions: first it irradiates a fluorescent sample, and, second, it starts an electronic timing sequence. The basis of the 'time-correlated single photon counting' method is the timing of the arrival of the first fluorescent photon at a sensitive photomultiplier relative to an arbitrary time zero, set by the firing of the flash lamp. This time difference is caused in part by the time taken for the pulses to travel through cables and electronics and by the time taken for the molecule to re-emit an absorbed photon. The time before re-emission varies with different excited states and defines their decay function. A time-to-pulse height converter provides a voltage pulse that is proportional to the time difference. This voltage is subsequently fed to a multichannel analyser operating in the pulse height analysis mode, and is counted in an appropriate channel. Repetitive lamp pulsing (lamp frequencies up to 200 kHz) and photon collection result eventually in a histogram of voltages being obtained in the memory of the multichannel analyser which is a direct analogue of the fluorescence decay function. In this technique, the time resolution is set by the jitter in the photomultiplier transit time (typically 300 ps) and not by the photomultiplier rise time. Furthermore, in this technique, photomultiplier 'shot' noise has little effect on signal-to-noise ratio and sensitivity is set principally by the number of accumulated fluorescence counts, which can be very large. A schematic diagram of a 'nanosecond fluorometer' is shown in figure 7, and figure 8 (plate II) shows a typical record as it appears on the multichannel analyser. The fluorescence decay profiles show a molecule undergoing a rearrangement (to an intramolecular exciplex) after it has absorbed a photon of light. The trace which appears and disappears most rapidly, the one on the left, is the fluorescence of the original molecule and the other trace is the fluorescence of the rearranged molecule which is formed from it and which itself then fluoresces. The appearance of the new molecule coincides with the disappearance of the original molecule.

Fluorescence decay measurements have also been used to give information on intermolecular and intramolecular energy transfer, radiationless processes in molecular systems and atomic interactions, to provide fundamental data such as quantum yields and quenching efficiencies and even to find use in qualitative analysis.

Time-resolved emission spectroscopy in the nanosecond time scale is a technique whereby the emission spectrum of a sample is obtained at different

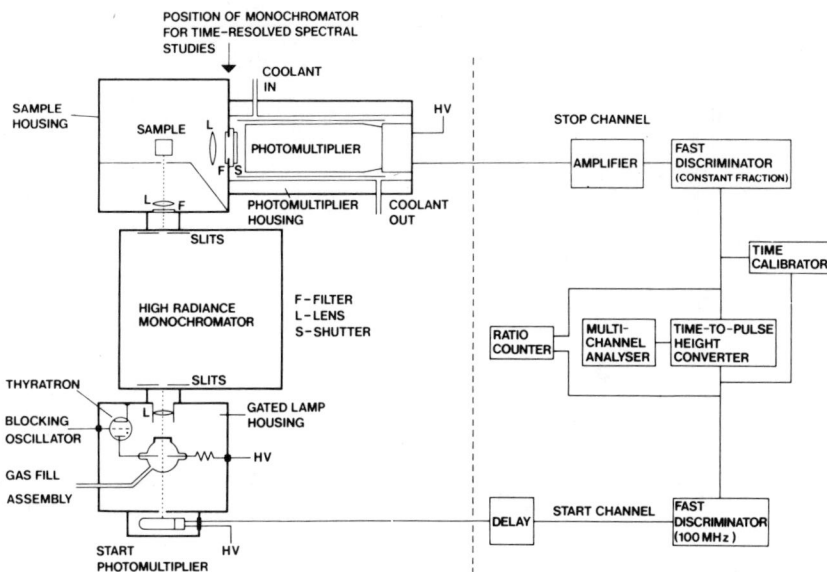

FIGURE 7. Optical and electronic system for determinations of nanosecond lifetimes and time-resolved emission spectra by time-correlated single photon counting.

times after excitation by photon counting during pre-selected time windows in a decay curve. Changes in the fluorescence spectrum in this time scale may result from solvent reorientation and interaction of the excited molecule causing relaxation of the excited state, electronic energy transfer from the initially excited donor to an acceptor or bimolecular interactions in the excited state such as excimer or exciplex formation. A dramatic example of a large spectral shift with time is shown in figure 9.

NANOSECOND FLASH PHOTOLYSIS

The discovery of pulsed lasers (particularly ruby and neodymium) provided a new source of excitation in flash photolysis which is not subject to the energy-duration characteristics of plasma flash lamps and which has made possible the extension of flash photolysis from the microsecond range into the nanosecond and picosecond regions. Lasers suitable for flash photolysis can be classed as follows:

1. Continuous gas lasers such as helium–neon, helium–cadmium, argon and krypton can be mode-locked to produce repetitive sub-nanosecond pulses of relatively low energy suitable mainly for emission studies.

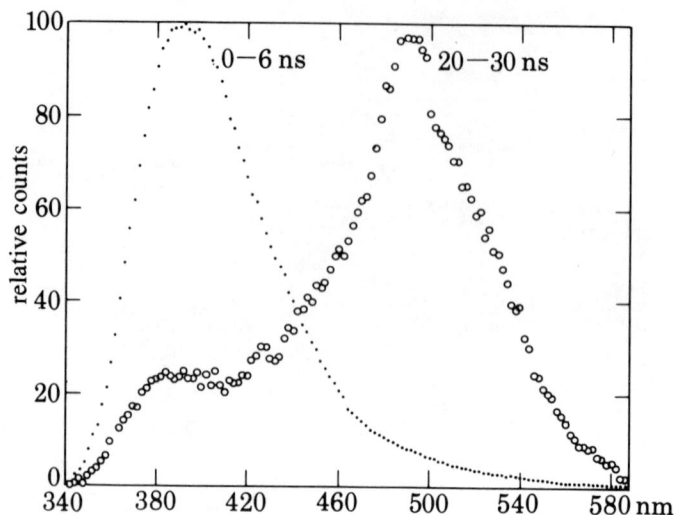

Figure 9.   Time-resolved fluorescence spectrum of a coumarin dye. Spectrum
lower wavelength peak recorded between 0 and 6 ns after excitation, spectrum
with higher wavelength peak recorded 20–30 ns after excitation.

2. Pulsed gas lasers such as nitrogen and hydrogen emit ultraviolet pulses of a
few nanoseconds width up to a megawatt in power. High power carbon dioxide
lasers emit in the infrared at 10.6 μm. These lasers are suitable for single-shot
nanosecond flash photolysis.

3. Liquid lasers including dye lasers emit in the ultraviolet to the infrared and are
tunable over a wide range. If flash-lamp pumped, they have a duration of about
0.5 μs for a 50 mJ pulse (pumped with a 100 J flashlamp). They may be
mode-locked to produce sub-nanosecond pulses.

4. Solid-state lasers were the first operational lasers and the first to be used for
nanosecond flash photolysis. They are still the most useful for this purpose. The
principal ones are the ruby laser, which emits at 694 nm and may be
frequency-doubled to 347 nm and the neodymium laser which emits at 1060 nm
and may be frequency-converted to 530, 354 and 265nm. The output consists
of a series of spikes of total duration equal to that of the pumping flash, i.e. about
1 ms but this is Q-switched, e.g. by a rotating mirror, Pockels cell or saturable
absorber, to produce a pulse which is typically between 2 and 20 ns. These lasers
may be mode-locked and neodymium is usually found to be the most
reproducible for this purpose.

   Given a laser pulse, monochromatic and of sufficient energy to produce a
measurable photochemical effect, the problem which remains is to record the
time-resolved spectrum immediately after excitation and for some period
afterwards. As in conventional flash photolysis, two monitoring methods are

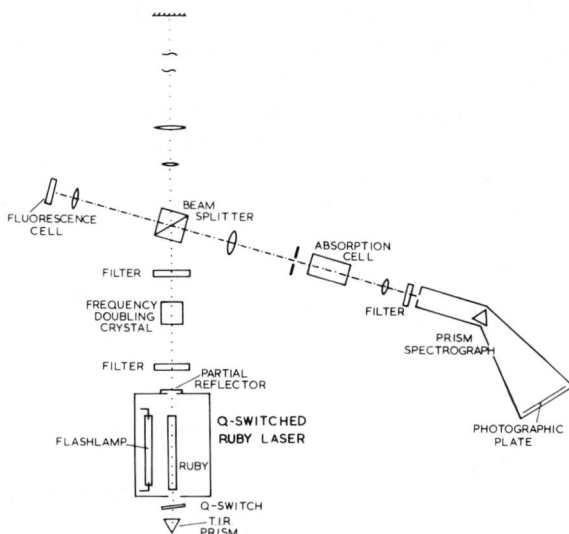

FIGURE 10. The experimental arrangement for laser flash photolysis and spectroscopy in the nanosecond region.

available in principle, (a) photometric recording over a narrow wavelength band at all times and (b) photographic recording over a wide range of wavelengths in a narrow time interval after a delay. The former is sensitive and more convenient for kinetic work and is limited by the response time of the detection system to about 5 ns. The latter is more convenient for a preliminary survey or for accurate spectroscopic work and is essential for most sub-nanosecond work.

To take full advantage of the short-lived laser pulses in flash photolysis, spectroscopic recording in the nanosecond region requires a monitoring flash with a duration comparable with that of the laser pulse and an output continuous over a useful part of the visible and near ultraviolet region. The second flash must be triggered by a reliable means a few nanoseconds after the laser flash with a variable delay reproducible to 1 ns. How this was achieved is illustrated in figure 10. First, the laser pulse is divided into two parts by a beam-splitter. The reflected part passes directly to the absorption cell and produces the effect to be studied. The second part is used to record the spectrum after a pre-determined delay time. The delay is easily introduced by using the finite velocity of light and causing the monitoring part of the laser beam to be optically delayed. Light travels 30 cm in one nanosecond so the adjustment of the delay path is easy. The second problem, that of converting the monochromatic wavelength of the laser pulse into a continuum, was solved by exciting the fluorescence of a solution of an organic scintillator, such as tetraphenyl butadiene, which has a lifetime so short that no significant time delay was introduced and which emitted a broad band of wavelengths. The monitoring light now passes through the absorption cell and into a spectrograph. An example of the records which were obtained in this way is shown in figure 11, plate II. The first product formed of this aromatic molecule after the absorption of light is that same molecule in its singlet excited state which subsequently decays to its triplet state.

Kinetic recording of nanosecond events is very similar to that used for conventional microsecond work except that it is often more convenient to use a collinear excitation and monitoring arrangement (with the transmitted laser light removed by an absorption filter), and an intense monitoring source (usually a long-lived flash or pulsed arc source) is needed for adequate signal-to-noise ratios. The technique has been applied to studies of excited state absorptions of polycyclic hydrocarbons in the gas phase using a 1 metre absorption cell and allowed direct observations to be made of vibrational relaxation following intersystem crossing and singlet–singlet absorption. Other examples of nanosecond flash photolysis include detailed studies on the photochemistry of the quinones, particularly those involved in the primary electron transfer processes of photosynthesis, the rapid quenching of excited states by oxygen and solvated electron kinetics.

PICOSECOND FLASH PHOTOLYSIS

In spite of having extended the time resolution of flash photolysis from microseconds to nanoseconds with pulsed lasers, there were still many chemical events which escaped detection and study because they were too fast. It happens that, as one seeks time resolution below one nanosecond, several new experimental problems appear. First, sub-nanosecond pulses necessitate mode-locking of a laser. Second, the recording of transient events in real-time using photomultipliers and oscilloscopes ceases to be feasible and, even if these instruments were capable of picosecond time resolution, the signal-to-noise ratio obtainable with continuous-light monitoring sources would be inadequate for most single pulse experiments. Third, the synchronization of two separate light sources, such as is done conventionally in microsecond flash photolysis experiments, is not possible in picosecond intervals of time.

Real-time recording in the picosecond region is possible by use of a streak camera. This photoelectronic device can record a decay profile in a single shot with a time resolution of about 3 ps. With image-intensified vidicon detection and recording by means of an optical multichannel analyser (OMA), the sensitivity and precision of the arrangement is suitable for qualitative kinetic measurements for light emission experiments such as fluorescence lifetime measurements. The arrangement of a solid state laser producing single picosecond pulses at 530 nm and a streak camera for lifetime studies on fluorescence is shown in figure 12. Studies of time-resolved primary energy transfer and light harvesting processes in photosynthesis with a red alga, *Porphyridium cruentum*, has shown how fluorescence decay profiles recorded at

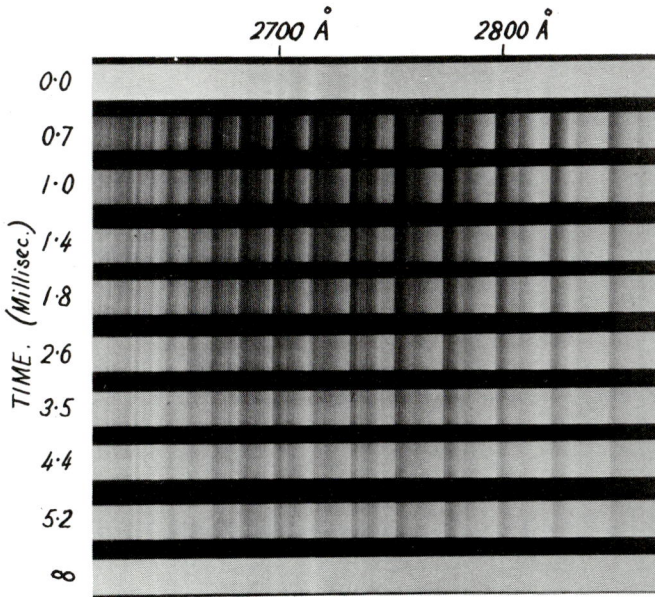

FIGURE 3. Sequence of spectra recorded by flash spectroscopy. The reaction vessel contained a mixture of chlorine and oxygen. The first spectrum was photographed before the photolysis flash and subsequent spectra at increasing delay times shown. The band spectrum is characteristic of the radical ClO.

PLATE II (PORTER & WEST)

excitation
of the
compound

OH

NHPh

in dilute
solution
in benzene

monomer decay   exciplex decay $\Big\}$ $\lambda_{\text{excit.}} = 290$ nm

intensity →

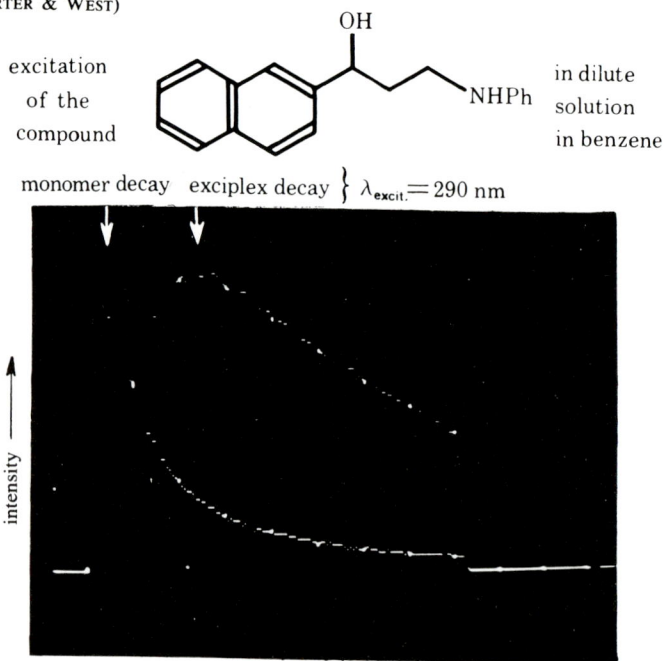

FIGURE 8. Record taken by the single-photon counting method of the internal rearrangement of an excited molecule. Naphthalene fluorescence is observed at 350 nm and exciplex fluorescence above 500 nm. The time interval between each bright spot is 10 ns.

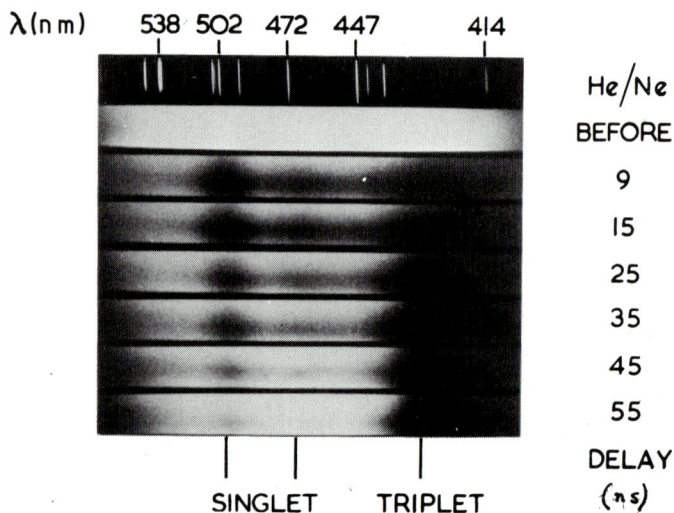

λ(nm)   538 502   472   447   414

He/Ne

BEFORE

9

15

25

35

45

55

DELAY
(n s)

SINGLET   TRIPLET

FIGURE 11. A typical record obtained with the arrangement shown in figure 10. The disappearance of the excited singlet state of triphenylene and appearance of the triplet state is observed over 50 nanoseconds. Delay time in nanoseconds.

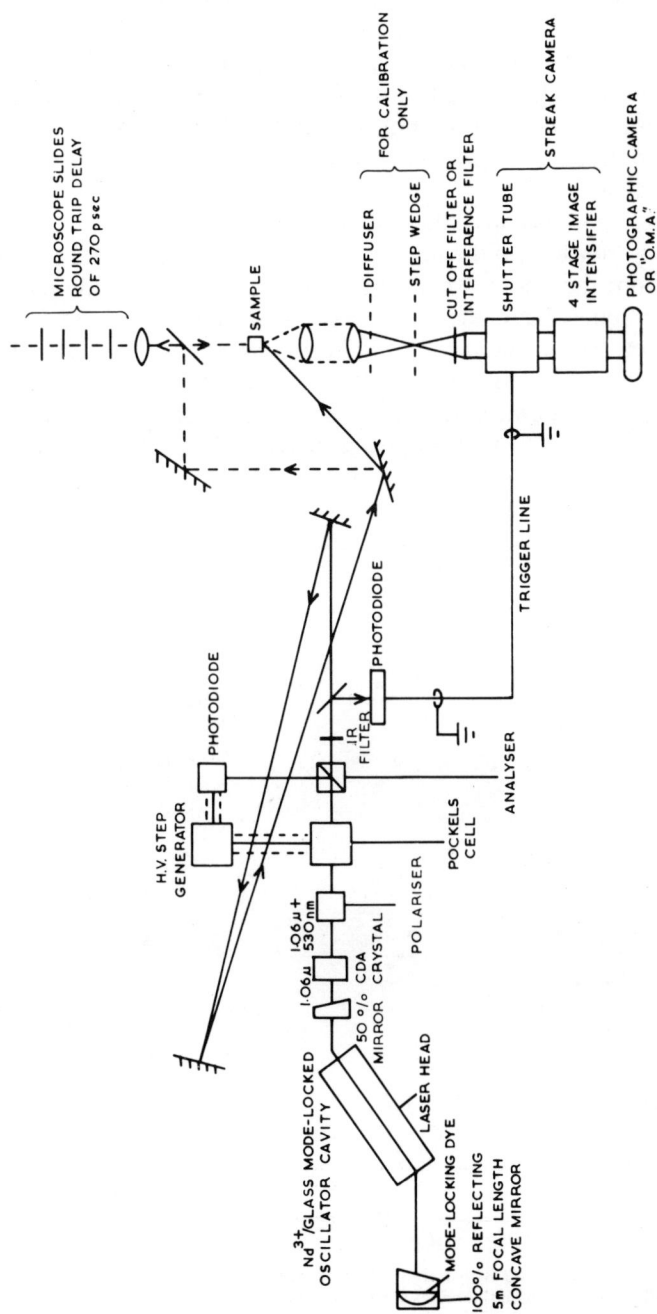

FIGURE 12. Arrangement for producing single picosecond pulses from a neodymium laser and for measurements of fluorescence lifetimes using a streak camera.

different wavelengths with this experimental arrangement can give a clear account of successive events of energy transfer. Typical results are illustrated in figure 13. This alga derives its red colour from auxiliary pigments which transfer energy eventually to chlorophyll-*a*. With 530 nm irradiation the pigment first excited is phycoerythrin (BPE). This transfers its energy to phycocyanin (RPC) which in turn transfers energy to allophycocyanin (APC) and thence, ultimately, the chlorophyll-*a* (CHL). The sequence of grow-in and decay of the pigment's fluorescence is clearly seen.

For absorption spectroscopy, the streak camera is unsuitable because for adequate signal-to-noise ratio the high intensity monitoring source necessitates high emission from the photocathode and consequent loss in time resolution. Absorption techniques have therefore been developed which resort to the original flash photolysis technique of monitoring using a second pulse. Even at the shortest time, a simple split-beam method for accurate synchronization of the exciting pulse with the monitoring pulse is available where the velocity of light determines the optical path by which the second part of the pulse is delayed by 0.3 mm/picosecond. With a little ingenuity, almost any type of time-resolved experiment can be carried out in this way. A single pulse from a mode-locked laser is split, by partial reflection, into two parts: the first part excites the sample and the second monitors the consequence of this excitation at a succession of

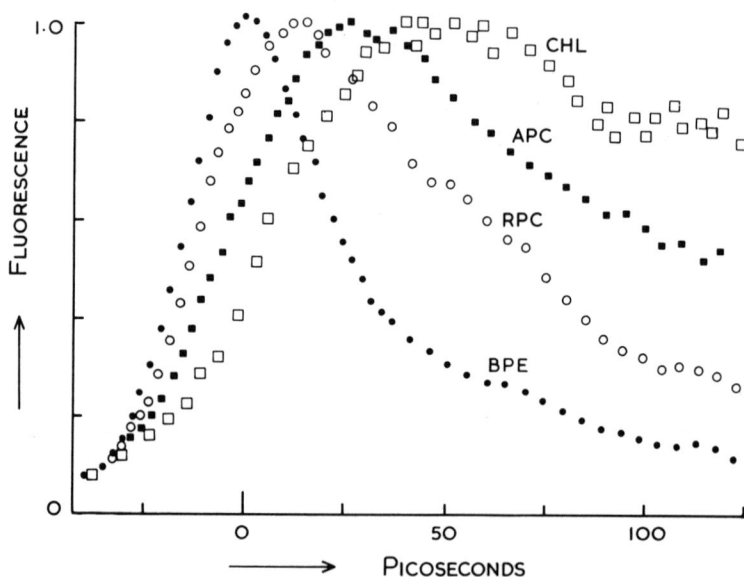

Figure 13. Energy transfer in red alga. Fluorescence of BPE at 578 nm decays by transfer in 70 ps to RPC which fluoresces at 640 nm. This transfers in 90 ps to APC which fluoresces at 660 nm. Finally this transfers in 120 ps to CHL which fluoresces at 685 nm and decays in 180 ps by transfer to the trap. (Carried out in collaboration with Dr. J. Barber and his group at Imperial College.)

times after excitation. Some examples of the ways in which the monitoring or probe pulse can be used are as follows:

1. Measure the transient absorbance at the wavelength of excitation.

2. Measure the absorbance at a wavelength different from the excitation wavelength. The wavelength of the probe pulse is changed by frequency multiplication, mixing, Raman shift, etc.

3. Measure the absorption spectrum over a range of wavelengths. The continuum necessary for this purpose is generated by focusing the probe pulse into a medium such as glass or water where self-focusing and self-phase-modulation broaden the line into a band of wavelengths which may extend over several thousand cm^{-1}.

4. Use the probe pulse to excite Raman scattering.

5. Use the probe pulse to cause two-photon absorption, the resultant state of which is detected, for example, by fluorescence. When no intermediate state is involved, two pulses must be absorbed simultaneously and this provides a method for measuring pulse widths. When an intermediate state of finite lifetime is formed by the exciting pulse, the method measures the lifetime of this state.

6. Use the probe pulse to open a shutter. The only shutter fast enough for picosecond work is one using the optical Kerr effect and this has been used to measure fluorescence lifetimes down to about 10 ps (and to photograph light in flight).

7. The probe pulse is mixed, in an optically nonlinear material, with light emission, e.g. fluorescence, excited by the photolysis pulse, and the intensity of the sum or difference frequency gives a measure of the fluorescence intensity at the delay time used.

The selection of a single pulse from a laser train is a very irreproducible procedure and to obtain even approximate quantitative measurements, a simultaneous comparison must be made with the pulse intensity. This may be done (i) by splitting the pulse further and measuring the energy of a known fraction of each pulse, (ii) by a double-beam experiment whereby only part of the beam contains the sample and (iii) by a multiple-beam experiment where different spatial sections of the same beam pass through different thicknesses of optical material (e.g. an echelon) and the whole sequence is recorded in one pulse.

This list is by no means exhaustive and, when one remembers that there is now a variety of solid-state lasers and dye lasers available for picosecond pulse generation, it is not surprising that, at present, there are as many variations of technique as there are workers in the field. Examples of chemical systems which have been studied by picosecond techniques are geminate recombination of iodine atoms in solution, hydrogen atom transfer in benzophenone, electron transfer from thiazine dyes, charge transfer interactions between anthracene and an aromatic amine, the decay kinetics of tetracene dianions and inorganic complexes.

As picosecond techniques allow molecular kinetic studies to be extended to shorter and shorter times, each new decade of time has revealed new phenomena as interesting and as numerous as the decade before it. In figure 14 some of the fastest and most fundamental molecular processes are shown with the regions in time in which they are most often found to occur.

Immediately a molecule has absorbed a quantum of light it is in a highly non-equilibrium situation in several respects and will begin to relax towards equilibrium by several routes, each with its characteristic rate. Even an isolated molecule will rotate and undergo internal redistribution of energy. In solution, it will undergo rotational diffusion, solvent shell relaxation, vibrational relaxation and emission. Internal conversion between electronic states, also possible in the isolated molecule, is again usually to be found in the picosecond region of times except for transitions which involve the ground state.

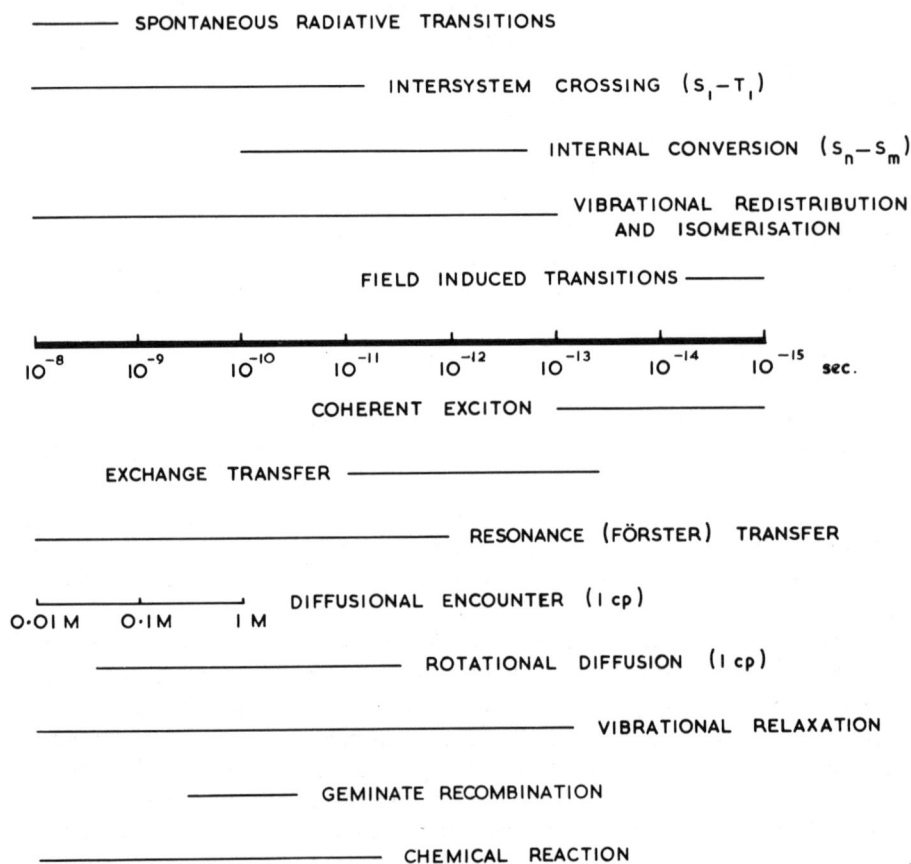

FIGURE 14.   Sub-nanosecond events in molecular physics and chemistry.

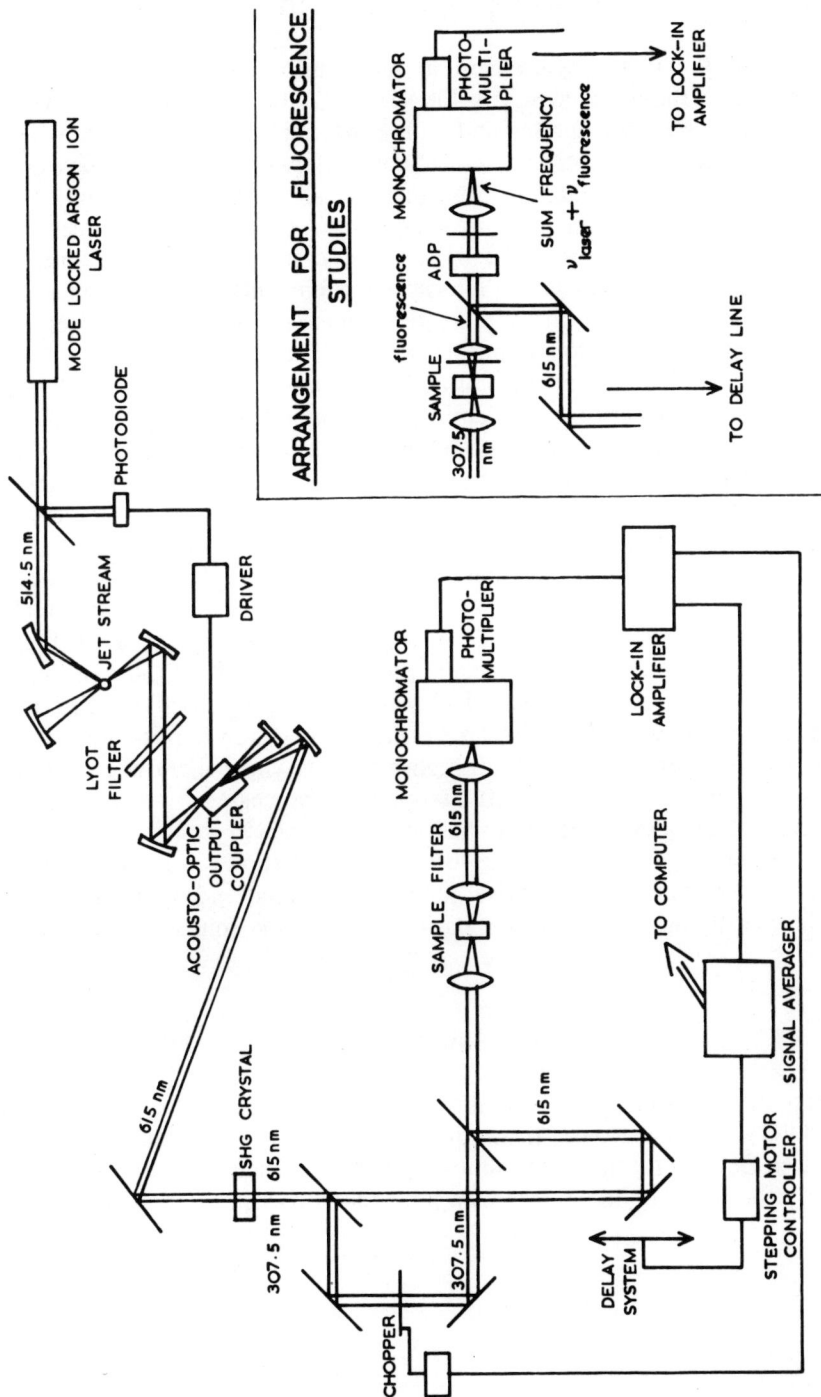

FIGURE 15. Experimental arrangements for sub-picosecond absorption and fluorescence spectroscopy.

SUB-PICOSECOND KINETICS AND FUTURE TRENDS

If we carry out a measurement in one femtosecond (10^{-15} second) the associated uncertainty in energy is 400 kJ/mole. Since this is comparable with the strength of a chemical bond and with a spectral band covering the whole of the visible region, chemistry and molecular physics may be said to have ended. Nevertheless, investigations of molecular events in the region of one or two picoseconds show that much of interest occurs in these very short times and an improvement in time resolution by at least a factor of ten, to about 100 femtoseconds, would be useful. Sub-picosecond events include intramolecular and solvent-induced vibrational relaxation, internal conversion between upper electronic states and exciton migration. For short laser pulses, large bandwidths are essential and therefore one turns to dye lasers. Pulses obtained from a mode-locked dye laser (e.g. rhodamine 6G) pumped continuously by an argon ion laser are of about 1 ps duration. By passing the laser pulse through a system with anomalous dispersion (e.g. a pair of diffraction gratings set at an angle so that the path traversed by the long wavelengths is greater than that traversed by shorter wavelengths), it is possible to reduce this pulse duration to about 0.3 ps.

The energy in pulses of this kind is insufficient for single pulse experiments. However, what one obtains is a uniform continuous pulse train of about 10^8 pulses per second and by use of an acousto-optical coupler (cavity dumper) this may be reduced to a manageable repetition rate of say 50 kHz with an attendant increase in pulse energy. The energy per pulse is small (say 3 nJ) but by signal averaging and long pulse trains, higher precision is possible than with single pulse experiments of much higher energy whilst the low-energy pulse eliminates the complexities of two-photon processes. An experimental arrangement using these principles has been developed in our laboratories by Dr. Graham Fleming and Dr. Godfrey Beddard and is shown diagrammatically in figure 15. Progress in short-pulse generation is rapid and within a few years kinetic measurements should be possible to the time limits of chemistry.

ACKNOWLEDGMENTS

Most of the later work described was carried out in the Davy Faraday Laboratory of the Royal Institution and we wish to thank the Science Research Council for their continuous support. We would also like to thank most warmly the many collaborators who have taken part in these developments.

THE INTRACELLULAR ELECTRODE:
25 YEARS OF RESEARCH IN CELLULAR ELECTROPHYSIOLOGY

P.F. BAKER, F.R.S.

Department of Physiology, King's College, London

INTRODUCTION

The electrical properties of membranes play a central role in the regulation of cell behaviour and function and nowhere are they more highly developed than in the nervous system. The quantitative study of these properties has been greatly facilitated by the development of techniques for inserting fine electrodes into cells. In the 1920's Osterhout and his colleagues showed that it was possible to insert electrodes into large plant cells; but the application of this technique to animal cells was not reported until 1939 when two Cambridge physiologists, A.L. Hodgkin and A.F. Huxley succeeded in introducing a fine capillary electrode into a giant axon from the squid. Since that time intracellular recording techniques have undergone continuous development and have been applied to an ever increasing number of problems. This account describes some of these developments and the results that have been obtained.

EXPERIMENTS ON LARGE CELLS

Background to the Hodgkin–Huxley (1939) Experiments

Towards the end of the eighteenth century it was already appreciated that electricity could be generated by certain biological systems—the so-called excitable cells: but the study of the minute electric currents involved had to wait the development of adequately sensitive methods of electrical recording. With the advent of such techniques they were soon applied to the nervous system notably by du Bois Reymond and Bernstein in Germany, by Lucas and E.D. Adrian in Cambridge and by Erlanger and Gasser in the USA; but a major limitation was that they could only detect the currents flowing in the medium outside the nerve cells and as most of their preparations were multicellular it was difficult to study the behaviour of an individual cell.

The next significant advance was made possible by the discovery of nerves of an extraordinarily large size. The British anatomist J.Z. Young found that the

175

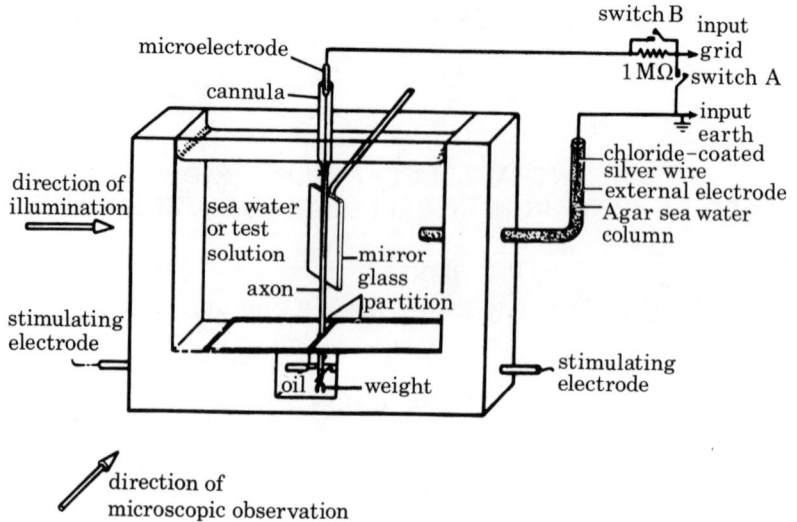

FIGURE 1. Apparatus used by Hodgkin, Huxley and Katz to introduce a fine axial electrode into a squid axon.

jet propulsion mechanism in squid is controlled by nerve cells, the axons of which can be up to 1 mm in diameter and many centimetres in length. Here was a large single nerve cell infinitely more suitable for experimental analysis than the minute nerves of mammals which rarely measure more than 1/100 mm in diameter, and Hodgkin and Huxley in Britain and at much the same time Cole and Curtis in the USA were able to exploit the size of the squid giant axon preparation to achieve the first intracellular recordings of electrical activity in an animal cell.

The technique used by Hodgkin and Huxley is shown diagrammatically in figure 1. A glass cannula was inserted into one end of the axon and a fine glass capillary less than one tenth of a millimetre in diameter (figure 2) introduced through the cannula into the interior of the axon. The glass capillary, filled with a conducting salt solution, established electrical contact with the interior of the cell, making possible for the first time the direct measurement of the difference in potential between the inside and outside of a cell. Their findings confirmed and extended earlier observations with external electrodes; in particular two facts emerged:

(1) they confirmed that in the absence of nervous activity a potential (*resting potential*) exists across the surface of the nerve cell, the interior of the axon being negative with respect to an electrode placed in the external solution;

(2) they found that when a nerve impulse is conducted along an axon the potential changes (*action potential*) and the interior of the cell becomes positive for a brief period: this swing in potential from inside negative to inside positive

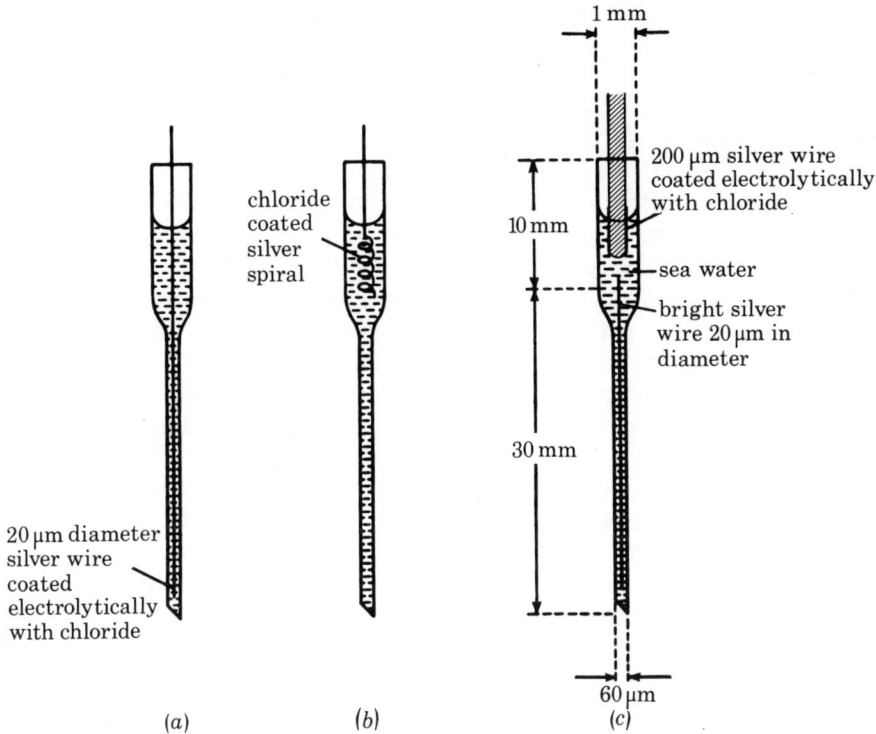

FIGURE 2.   Three designs of capillary electrode used by Hodgkin, Huxley and Katz.

and back again to inside negative was quite unexpected as previous workers had suspected that the membrane potential collapsed to zero during activity.

This work established the squid giant axon as one of the most valuable single cell preparations available to neurophysiologists interested in the mechanism of nervous conduction. Britain is particularly fortunate in having a species of squid, *Loligo forbesi*, which has unusually large axons even for squid and the Laboratory of the Marine Biological Association at Plymouth has become one of the main centres in the world for this type of research.

Much of the subsequent work on the squid axon has been directed towards understanding the mechanisms by which the resting and action potentials are generated.

The Resting Potential

Two important clues to the origin of the resting potential came from the chemical analysis of axoplasm and the sensitivity of the resting potential to the ionic composition of the medium in which the axon is immersed.

By gently squeezing squid axons Bear, Schmitt and Young were able to obtain samples of nerve cell protoplasm—axoplasm—virtually uncontaminated by extracellular fluid. Analysis of this axoplasm revealed striking differences in ionic composition when compared with squid blood. Potassium ions are present in higher concentrations in axoplasm whereas sodium, calcium, magnesium and chloride ions are all present in lower concentrations than in plasma. This distinctive intracellular environment is not unique to nerve but is found in virtually all cell types and is maintained at the expense of cellular energy by enzymes located in the surface membrane of the cell.

The resting potential is sensitive to the concentration of potassium ions in the medium bathing the axon, but is rather insensitive to the concentration of sodium. Increasing the potassium concentration reduces the resting potential and this effect is fully reversed on returing to the initial solution. Quantitatively there is a linear relation between the resting potential and the logarithm of the external potassium concentration.

Taken together these two sets of observations suggest that the resting potential may be generated by the tendency for K ions to diffuse out of the nerve cell from a region of higher concentration to one of lower concentration. If K ions can cross the membrane more easily than other ions, the outward diffusion of potassium should establish a potential whose size would be determined solely by the steepness of the gradient of K ions. Put in a slightly different way, the battery

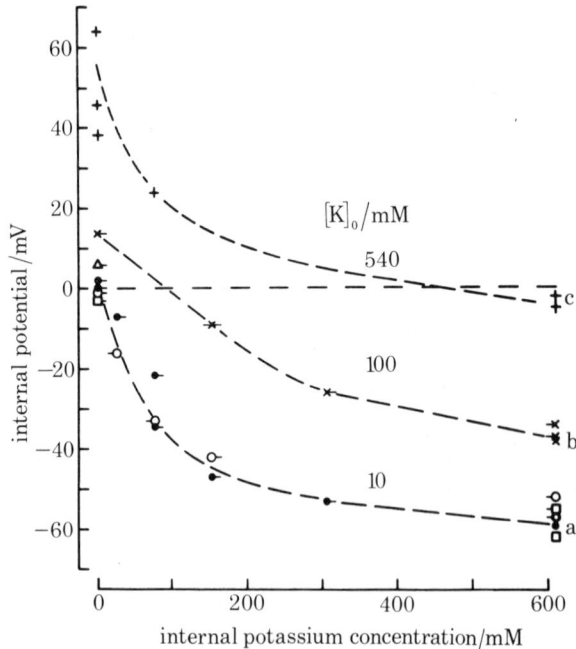

FIGURE 3.   Dependence of the resting potential of a perfused squid axon on the steepness and direction of the gradient of potassium ions. (From Baker, Hodgkin and Shaw.)

for generating the resting potential is the difference in activity of K ions in the two sides of the membrane.

One of the most dramatic proofs of this proposition was obtained in the early 1960's by Baker, Hodgkin and Shaw working with perfused squid axons. They showed that after squeezing the protoplasm out of a squid axon, the flattened sheath can be reinflated with an artificial protoplasm and, provided the solution used is rich in K ions and has roughly the same osmotic pressure and pH as normal axoplasm, the perfused axon will conduct nerve impulses. This observation provided strong evidence that the bulk of the protoplasm is not necessary for nervous conduction and reinforced the view that the source of energy for the generation of potentials is the ion gradients. Figure 3 shows that the magnitude and sign of the resting potential is determined solely by the steepness and direction of the K ion gradient. With a high K concentration inside the axon and a low K concentration outside, the resting potential is of the normal size and the interior of the axon is negative. But when these conditions are reversed—with high K outside and low K inside—the interior of the axon becomes positive and with equal concentrations of K on the two sides of the membrane the potential is close to zero.

The Action Potential

The striking feature of the action potential is that the interior of the axon transiently becomes positive. The source of this positive potential is another ion gradient, that of sodium. Sodium ions are present in high concentration in the external solution and replacement of them by organic cations such as choline abolishes the action potential and with it the ability of nerves to conduct impulses. By varying the Na concentration inside and outside an axon it can be shown that the extent by which the interior of the axon becomes positive is determined by the steepness of the inward gradient of Na ions (figure 4).

The upshot of these investigations was a rather simple model for the action potential based on the idea that the permeability of the nerve membrane to ions is not a fixed property but can change. Thus, at rest, the membrane is selectively permeable to K ions, but during the passage of an impulse, for a brief period, the membrane becomes selectively permeable to Na ions. Because of the direction of the ion gradients, this transient alteration in ion selectivity results in the potential changing from inside negative to inside positive and back again to inside negative.

The Voltage Clamp

The obvious question is: What causes the selectivity properties of the nerve cell membrane to change? In order to answer this question, techniques more sophisticated than the simple intracellular electrode were required. Instead of the internal potential swinging from negative to positive and back again in an

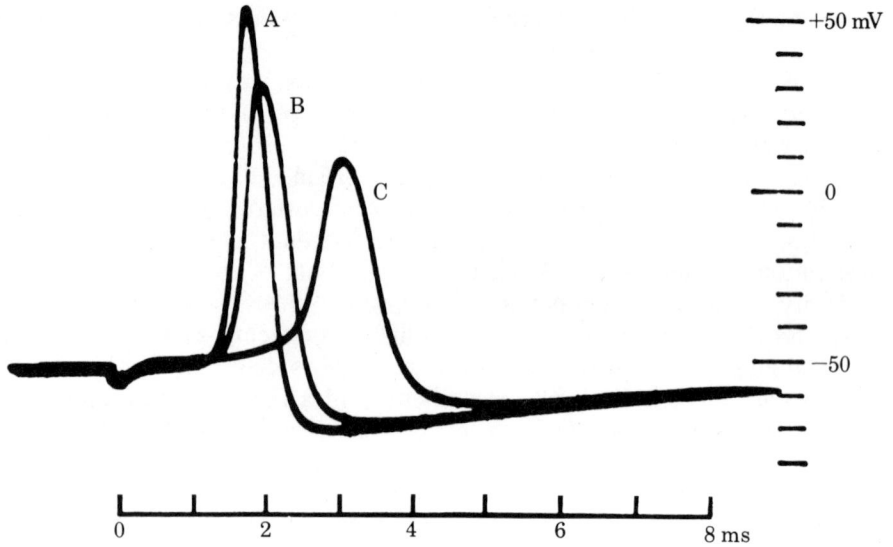

FIGURE 4. Intracellularly recorded action potential showing that the extent to which the potential 'overshoots' zero and becomes positive depends on the steepness of the gradient of sodium ions. In this experiment the axon was perfused and A, B and C reflect an increasing internal Na concentration. The records were obtained in the order B, A, C. (From Baker, Hodgkin and Shaw.)

uncontrolled manner, a method was needed for measuring the ionic current flowing across the membrane at fixed values of membrane potential. This was precisely the achievement of the voltage clamp technique introduced by Cole and Marmont in the USA and developed by Hodgkin, Huxley and Katz in Britain. By inserting two electrodes into the interior of the axon, one for measuring potential and the other for passing current, they were able to set the membrane potential at any predetermined level and hold it there while monitoring the current flowing across the axon membrane. A diagram of the voltage clamp electrode used by Hodgkin, Huxley and Katz and photographs of an actual electrode inside an axon are shown in figure 5, plate I.

Their key finding was that the transient increase in sodium permeability is initiated by a reduction in the transmembrane potential (depolarization). It is important to grasp the significance of this finding to the problem of nervous conduction. The interior of the resting nerve is negative. Anything that reduces this negative potential may initiate an action potential during which the interior of the axon will become positive. The region of local positive potential so created acts as a battery causing currents to flow along the axon into the neighbouring negative regions. As these local currents leave the axon to complete the circuit in the external medium, they serve to reduce the membrane potential and so set in motion the alterations in membrane permeability that in turn give rise to an action potential. In this way an action potential initiated at one end of a nerve is conducted without decrement to the other.

The voltage-clamp technique also revealed that in addition to increasing the sodium permeability depolarization has two other effects of particular relevance to nervous conduction. These are (1) that the increase in sodium permeability is only transient even in response to maintained depolarization and (2) that depolarization brings about a slower maintained rise in the permeability to potassium which serves to augment the resting potential mechanism in restoring the membrane potential to its initial value after passage of an action potential. The transient nature of the rise in sodium permeability is central to conduction. Without some means of terminating the increase in sodium permeability the nerve membrane would remain depolarized and conduction cease.

From a detailed analysis of the kinetics and potential-dependence of these changes in conductance, Hodgkin and Huxley in 1952 formulated a quantitative model of the conducted action potential which after 25 years is still the most successful treatment available.

Local Anaesthetics and other Agents

The understanding of nervous conduction provided by this work has given remarkable insight into the mechanism of action of many natural toxins and clinically useful agents that interfere with nervous conduction. From what has been said already, it is clear that the sodium permeability system is essential for normal conduction and anything that blocks it would be expected to block impulse traffic along a nerve, i.e. it would act as a local anaesthetic. Interference with the depolarization-induced increase in potassium permeability would have a less dramatic effect as this only serves to augment the resting potential mechanism. In the presence of blockers of the voltage-dependent K permeability, repolarization of the membrane will be slower and the action potential prolonged but conduction will continue.

Examination of many toxins of natural origin and clinically useful drugs has revealed a range of activities that can be loosely grouped under three headings:

(1) those that block the voltage-sensitive sodium channels

(2) those that interfere with the turning off or inactivation of the sodium channels

(3) those that interfere with the voltage-sensitive potassium channels.

As expected, substances in the first category are potent local anaesthetics blocking conduction with little or no change in resting potential. Notable examples are the puffer fish poison tetrodotoxin (TTX), the paralytic shellfish poison saxitoxin (STX) and a number of medically useful local anaesthetics for instance, xylocaine. Substances in the second category tend to depolarize the cell producing at first spontaneous electrical activity and eventually blockage of conduction. Examples are the veratrum alkaloids and the insecticide DDT. Substances in the third category prolong the action potential but do not block conduction. They include a number of K-like ions such as caesium and rubidium as well as a variety of quaternary ammonium compounds such as tetraethyl ammonium (TEA) and 4-amino pyridine. A number of substances have more than one action—particularly notable are the general anaesthetics which tend to affect a range of membrane processes possibly because their site of action is the membrane lipid in which both permeability and transport systems are embedded.

Towards Isolation of Voltage-sensitive Channels

Not only has the quantitative understanding of nervous conduction helped explain the action of a number of drugs and toxins, these agents have also been of great value in dissecting the conduction process still further. Thus the fact that blockage of the Na channels does not interfere with opening of the K channels and *vice versa* strongly suggests that the two sets of channels are independent and that the temporal changes in permeability properties of the membrane from K selective to Na selective and back again to K selective does not result from alterations in the properties of a single channel.

Another use to which toxins can be put is to determine the number of channels in a particular cell. In order to act, the toxin must first interact with the membrane and by using radioactively labelled toxin it has proved possible to count the number of toxin binding sites and from this gain an upper estimate for the number of channels in the membrane. Using labelled TTX and STX with a variety of excitable cells Ritchie and others have shown that the number of Na channels varies from about 5–$500/\mu m^2$ of membrane. This specific interaction between toxin and channel provides the most hopeful method for isolating voltage sensitive channels in a form amenable to chemical and physico-chemical analysis. To date, only partial purification has been achieved but it seems likely that this will be an area of great activity during the next 25 years. At this stage, the available data suggest that the sodium channel may be a protein large enough to span the thickness of the membrane.

In the absence of ways of isolating the voltage-sensitive channels, much recent activity has been directed towards electrical methods for studying the earliest events in opening the channels. It has always seemed likely that the exquisite voltage sensitivity of the sodium and potassium channels must result from the movement of charged groups within the membrane and many workers have sought evidence of such charged 'gating' particles. Recently great excitement has been aroused by the reports of Armstrong and Bezanilla in the USA, and Keynes and Rojas, and Meves in Britain that the opening of the sodium channel is associated with an asymmetrical movement of charge. If future work confirms that these 'asymmetry currents' reflect charge movements specifically associated with the voltage-sensitive permeability processes and not simply non-specific effects resulting from alterations in the electric field across the cell membrane, it should soon be possible to take the Hodgkin–Huxley analysis to a higher degree of resolution making possible a choice between at least some of the different physical models all of which are compatible with the Hodgkin–Huxley formulation.

EXPERIMENTS ON SMALLER CELLS

The Development of Microelectrodes

To what extent are data obtained on the squid axon applicable to other excitable cells? This is clearly a crucial question and its answer required the development of much smaller electrodes.

Graham and Gerard working in the USA in the forties first tried to use glass microcapillary tubing (tip diameter 2–5 μm) filled with salt solution to measure intracellular potentials. The technique used was to thrust the electrode tip directly into the cell. Unfortunately the cells suffered appreciable damage, but in 1949 Ling and Gerard introduced a significant improvement, the microelectrode which had a tip diameter of about 0.5 μm, and could be inserted into striated muscle cells with only minimal damage (figure 6). Subsequent workers have used electrodes with even finer tips in order to penetrate an ever increasing number of smaller cells. Provided the electrode tip is fine enough the cell membrane seems to seal around the electrode, minimizing damage.

Hodgkin and Nastuk in 1950 were the first to use microelectrodes to record rapidly changing potentials. They filled the electrodes with 3M KCl to give an electrical resistance of between 5 and 10 MΩ and in their recording system used cathode screened cathode followers. Since this early work with microelectrodes, though the electronics has changed dramatically, the microelectrode has

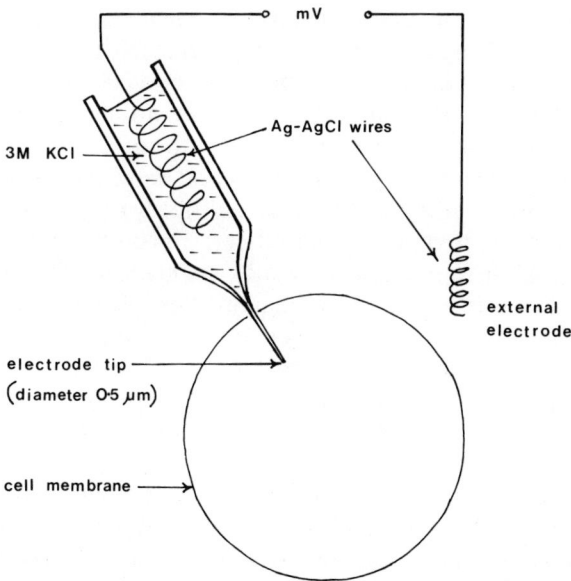

FIGURE 6. Diagrammatic representation of a microelectrode inserted into a cell. Figure 14, plate VIII shows a microelectrode inside a snail neurone.

remained virtually unchanged (figure 6) and has been widely used in cellular electrophysiology to investigate, among much else, neurones of the central nervous system, retinal receptors, the neuromuscular junction and synapses, the cardiac action potential and its relation to the electrocardiogram and the electrical properties of striated and smooth muscle. In this development it has become commonplace to insert not one but two and even three microelectrodes into a single cell making possible control of the membrane potential and extension of the voltage-clamp technique to a range of physiologically important small cells.

Microelectrodes in Muscle

Once it became possible to record intracellularly from a wide variety of cells, it also became clear that many of the principles enunciated in squid are rather generally applicable although individual tissues often have special electrical characteristics that can be related to particular physiological functions. A rather good example is the heart. The important physiological characteristics of the heart are its inherent rhythm which can be varied in response to suitable stimuli, its variable force of contraction and its resistance to tetanization. All three have their basis in membrane properties that have become accessible to analysis by microelectrode techniques. The cardiac action potential lasts about 200 times longer than that of a squid axon yet the work of Weidmann and Reuter in Switzerland, of Trautwein in Germany, of Hutter and Noble in Britain and of many others has shown that underlying the cardiac action potential are time and voltage-dependent changes in membrane permeability analogous to those studied in squid axons. The prolongation of the action potential, which serves to protect the heart against tetanization, comes about through greater separation in time of the inward sodium current and the outward potassium current. The delay in the onset of the outward potassium current is remarkable and not fully understood; but its properties of slow onset and decay are precisely those needed to explain the inherent rhythmicity of the heart. In addition throughout much of the action potential there is a small inward current carried by calcium ions. As muscular contraction is activated by calcium, this inward Ca current serves to couple excitation at the cell surface to contraction. At least one way of altering the force of contraction of the heart seems to be to alter the size of this inward calcium current. Thus stimulation of the sympathetic nerves to the heart increases the force with which the heart beats apparently because the noradrenaline released on nervous stimulation acts on the heart to make more calcium channels available for carrying Ca into the contractile apparatus.

A major complication in the study of the electrical properties of muscle is that the surface membrane is not smooth as it is in nerve but is usually infolded, often in a very regular manner. These infoldings serve to increase the membrane area and to facilitate transmission of the electrical signal to the contractile apparatus. Once the surface membrane becomes specialized it is not difficult to

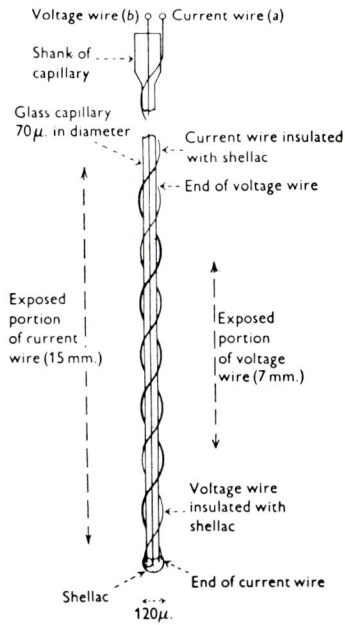

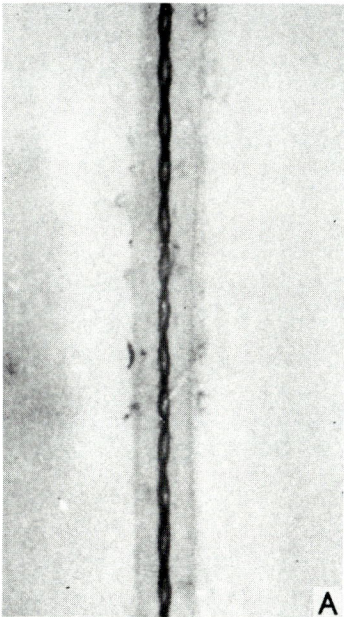

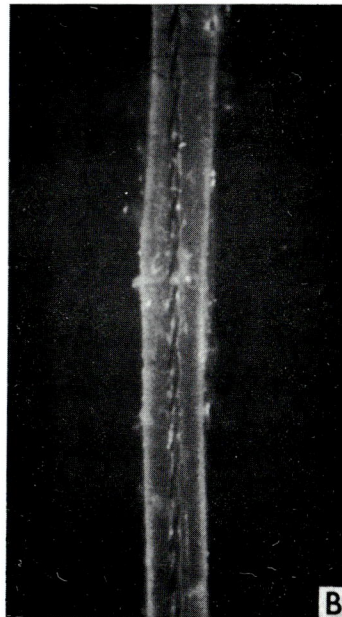

FIGURE 5. Upper. Diagram of the arrangement of current and voltage wires in the so-called 'double-spiral'
voltage clamp electrode of Hodgkin, Huxley and Katz.
Lower (A, B). Photographs of one of the original 'double-spiral' voltage clamp electrodes inside a squid axon.

Plate II (Baker)

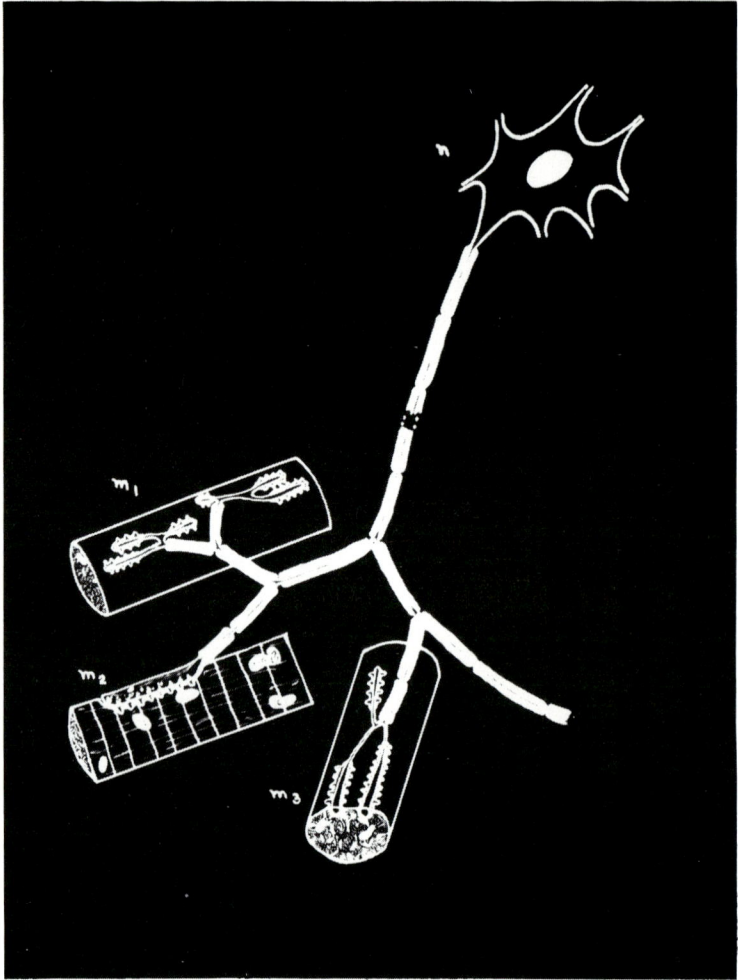

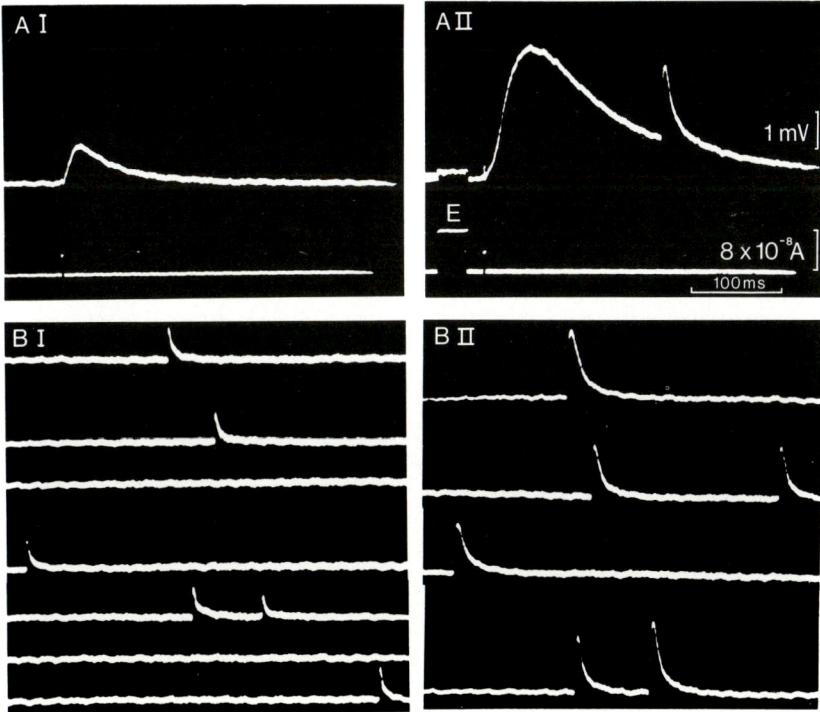

FIGURE 8. Miniature end-plate potentials (in B), showing the potentiating effect of a cholinesterase inhibitor (in II). In A, the effect of applied acetylcholine and its potentiation by the anti-esterase (AII) are shown at the same end-plate. (From Miledi.)

PLATE IV (BAKER)

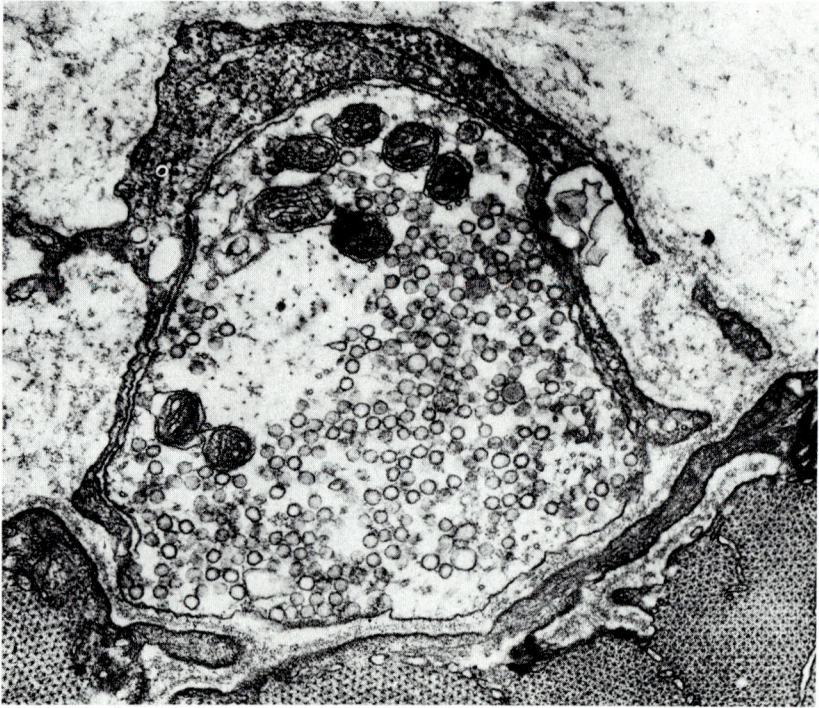

FIGURE 9.   Electron micrograph of a frog neuromuscular junction. The circular profiles are presumed to be vesicles that contain the quantal packets of acetylcholine. (From Miledi.)

FIGURE 10.   Squid giant synapse. The presynaptic axon has been injected with fast green dye to show the terminal synaptic branches in contact with the post-synaptic giant axons. (From Miledi.)

PLATE VI (BAKER)

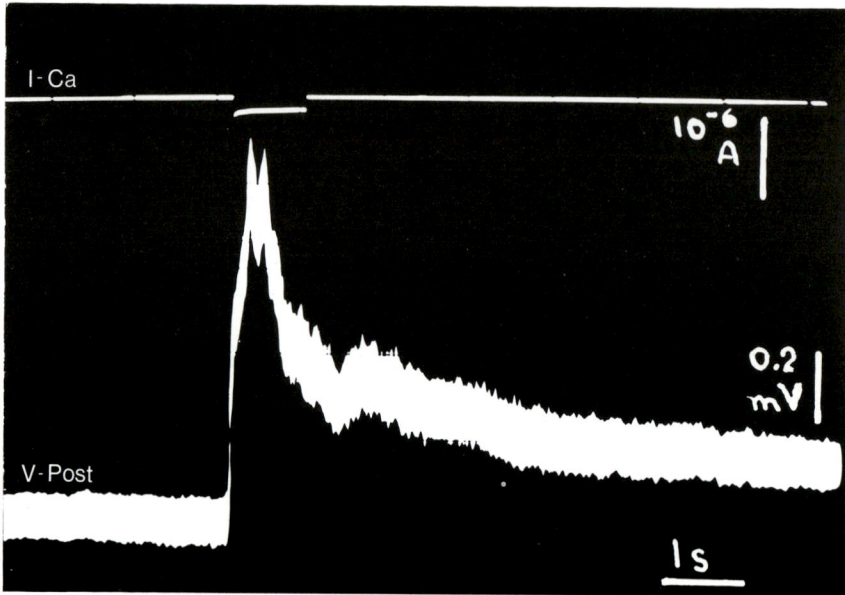

FIGURE 11. Transmitter release induced by direct ionophoretic injection of calcium into the presynaptic terminal of the squid giant synapse. The maximum rate of release corresponds to about 10^5 quanta per second. (From Miledi.)

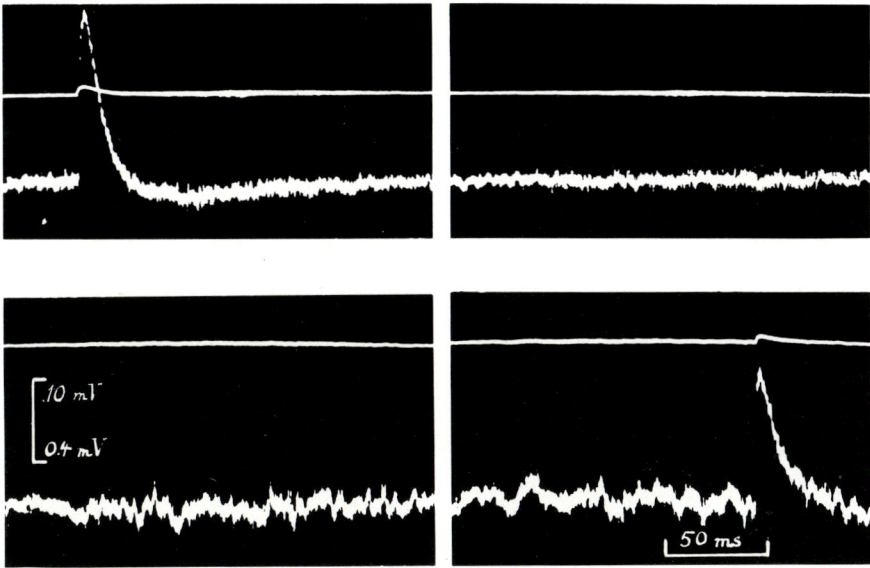

Figure 12. A statistical approach to the kinetics of molecular acetylcholine action. Each block shows simultaneous recording of membrane potential with a low-gain, direct-coupled (10 mV scale, upper trace) and high-gain condenser-coupled (0.4 mV, lower trace) channel. In the bottom row, a steady dose of acetylcholine was applied which depolarized and at the same time caused increased membrane voltage fluctuations ('acetylcholine noise'). (From Katz and Miledi.)

Plate VIII (Baker)

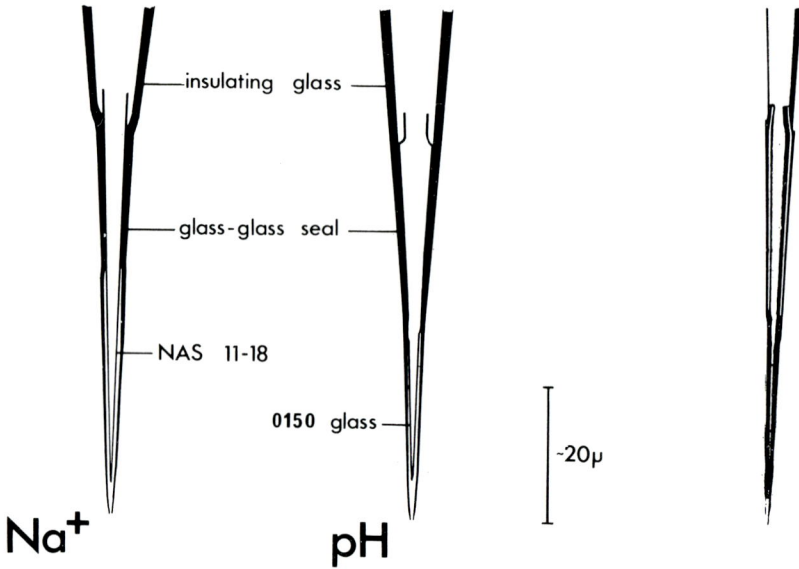

insulating glass

glass-glass seal

NAS 11-18

0150 glass

~20μ

Na⁺

pH

FIGURE 13.   Ion-selective microelectrodes. At the left diagrammatic representation of the sharp end of the electrode showing the recessed ion-selective tip. Only the extreme tip needs to penetrate the cell. At the right is a photograph of the tip of a pH-sensitive microelectrode. (From R.C. Thomas.)

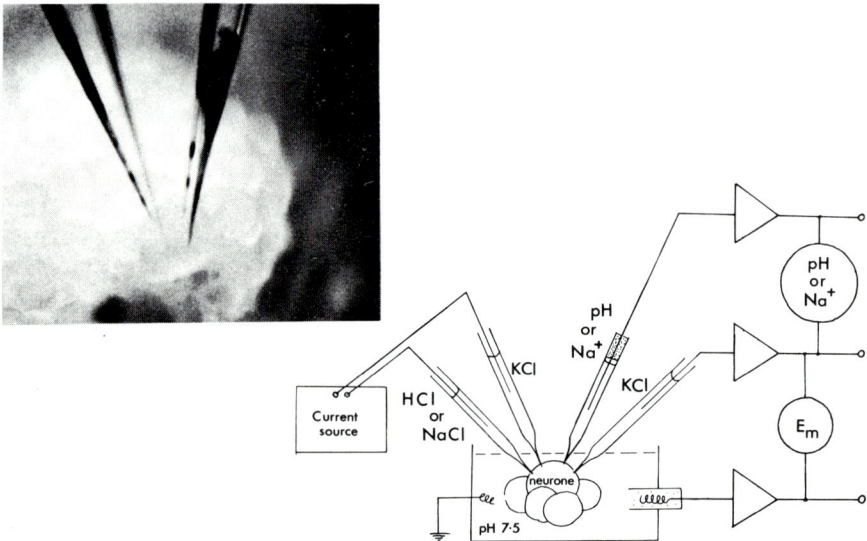

FIGURE 14.   Four microelectrodes in one snail neurone. The set up allows injection of H^+ or Na^+ into the cell whilst monitoring H^+ or Ha^+ with a suitable ion-selective microelectrode and conventional potential-sensitive microelectrode. (From R.C. Thomas.)

imagine that certain permeability properties may become restricted to particular regions of membrane. This seems to happen in muscle; but two rather more dramatic examples of such segregation within a single cell can be found in salt transporting epithelia and in the unicellular organism *Paramecium*.

Permeability and Transport across Epithelia

The physiological function of most salt transporting epithelia, such as those of the kidney and small intestine, is to move sodium, chloride and water from one side of a sheet of cells to the other. The basic mechanisms used seem to be essentially those we have met already—selectively permeable membranes and ion pumps; but the special trick in epithelia is that these are segregated within the cell. Thus Ussing and his co-workers in Copenhagen showed that the side from which Na is transported is selectively permeable to Na ions allowing these ions to enter the cell, and the opposite face of the cell is rich in Na pumps which expel the sodium from the cell into the medium bathing the other side of the epithelium. The face of the epithelium that is rich in sodium pumps has a low sodium permeability but is permeable to K ions which tend to leak from the cell only to be pumped back in exchange for sodium. The net effect of this organization is the vectorial flow of sodium—followed by chloride and water—across the epithelium. Anything interfering with these flows will tend to slow down salt and water transport—effects which in the kidney could produce an increased urine flow or diuresis and in the gut reduced fluid reabsorption or diarrhoea.

Permeability and Behaviour

The unicellular organism *Paramecium* provides a beautiful example of the role of membrane properties in the control of cell behaviour. When a *Paramecium* bumps into an object the direction of beat of its cilia reverses and the animal swims backwards rotating slightly. After a short while, the direction of ciliary beat returns to normal and the animal moves forward again but now on a somewhat different track. This pattern of behaviour only results from mechanical stimulation of the front end of the animal. A similar stimulus applied at the rear end only serves to speed up forward motion. Kinosita and Naito in Japan and Eckert and his co-workers in the USA have used microelectrodes to analyse the membrane events underlying this behaviour. They found that mechanical stimulation of the two ends of a *Paramecium* produce quite different membrane responses. Deformation of the anterior end causes a depolarization whereas the same stimulus applied to the posterior end causes hyperpolarization. These changes can be related to the behaviour of the animal. Hyperpolarization speeds up the ciliary beat; but depolarization opens up channels through which Ca ions enter the organism and this entry of Ca can be shown to cause reversal of the ciliary beat. The Ca that enters seems to be bound quite rapidly and as its concentration falls, the cilia return to their original direction of beat.

These observations on *Paramecium* provide insight into the complex ways in which the behaviour of a cell may depend on the permeability properties of its surface membrane, and with *Paramecium* it is not hard to see how alterations in permeability properties may affect behaviour. Such alterations might come about because of interaction with substances in the animals environment or because of changes in the genetic make-up of the animal leading to absence of components of the permeability system. Mutant *Paramecia* that lack the avoiding reaction, have been isolated and the defect has been shown to result from absence of calcium channels.

It seems likely that a number of instances of abnormal function in higher organisms may also result from alterations in the number or distribution of permeability channels in cell membranes; but the experimental analysis of these systems usually poses many problems. However, one rather clear example seems to be myotonia, a congenital disease in which even mild exertion leads to prolonged electrical activity in skeletal muscle making it difficult for the individual to remain standing. Though incapacitating, this condition is not fatal and is found in goats as well as man. Using the goat as a model, R.H. Adrian and Bryant have shown that the only detectable defect in the muscles is the absence of a normal chloride-permeability system in the muscle cell membrane. In muscle both potassium and chloride contribute to the resting potential and the lack of one of these systems is adequate to account for the observed behaviour.

Another remarkable example of the interaction between permeability and cell behaviour occurs when an egg is fertilized. A major problem during fertilization is to ensure that the egg is fertilized by only one of the many sperm that may be present. Rothschild and Swann first drew attention to the existence of a mechanism that comes into action soon after fertilization of an egg and serves to reduce the chance of a second sperm effecting fertilization. This so called 'fast block' to polyspermy occurs before any obvious morphological change takes place, but has very recently been shown to coincide with a reduction in the membrane potential of the egg. By manipulating the potential, Laurinda Jaffe has shown that depolarization of unfertilized eggs makes them more difficult to fertilize and hyperpolarization of fertilized eggs renders them susceptible to fertilization by a second sperm. Although much is still uncertain, there seems little doubt that the fast block to polyspermy employs mechanisms that have much in common with the processes that later in development underlie excitability in nerves and muscles.

TRANSMISSION OF INFORMATION FROM ONE CELL TO ANOTHER

Introduction

So far in this discussion we have been concerned with the electrical properties of individual cells; but in multicellular organisms cells must communicate with each other. In general cells communicate in two ways. An electrical signal may pass directly from one cell to its neighbour or a chemical messenger may be used. This second possibility involves considerable specialization both on the part of one cell for producing and releasing the message and on the other for detecting and responding to it. Although the concept of chemical messengers, hormones and transmitter substances, has been with us for half a century, it is only in the last 25 years—largely through the availability of microelectrodes—that we have begun to understand the intimate details of how these substances are released and act.

Electrical Coupling between Cells

Electrical coupling between cells was first revealed in 1959 by Furshpan and Potter working at University College London. They used microelectrodes to study the electrical properties of the crayfish giant motor synapse and found that it behaved as a simple electrical rectifier, that is, it passed current freely in one direction but only with great difficulty in the other. Subsequent work has shown that such junctions are rather rare in the nervous system but common in other areas of the adult body particularly the heart, smooth muscle and epithelial structures such as the liver, salivary glands and pancreas. In most of these tissues the junctions differ from that of the crayfish giant synapse in one important respect: they permit current to flow equally well in either direction.

In the heart and smooth muscle these junctions serve to allow current to pass from cell to cell ensuring that the whole structure acts as a co-ordinated whole. A similar function exists in the epithelia of some lower animals but in many epithelia and embryonic structures there is no obvious electrical activity yet junctions of low electrical resistance are present. The properties of these junctions can be probed by introducing marker substances into one cell and following whether or not these markers cross the junction. For this purpose, microelectrodes can be converted into microinjectors by the simple expedient of applying a positive pressure to their open end. By filling the electrode with a fluorescent dye, the subsequent movement of the fluorescent molecules can be followed rather easily. These experiments show quite clearly that provided the marker molecule is small enough it can pass from cell to cell apparently through the low resistance junctions. Experiments with dyes of different molecular weight have shown that the junctional pathway excludes molecules of molecular weight greater than about 1500 daltons. It is an attractive possibility that these junctions may provide a means of effecting direct chemical communication between cells without interchange of macromolecules; but rather little is known at present about the physiological role of these junctions although this does not stop speculation that absence of intercellular communication might be one factor contributing to certain malignancies. What is clear however is that the junctions are labile. During development they can be broken and reformed and variations in the metabolic state of the cell can lead to dramatic changes in electrical coupling. From the work of Lowenstein and Rose in the USA and of Warner and co-workers in London the key regulatory factors seem to be the intracellular activities of calcium and hydrogen and the possibility exists that *in vivo* alterations in the intracellular activity of these ions may serve to switch on and off electrical coupling between cells.

Chemical Transmission

The microelectrode has been instrumental in facilitating our understanding of how cells communicate through the release of specific chemicals. One of the clearest examples of such chemical communication is the process by which a nerve impulse is transmitted from one cell to another in the nervous system or from a nerve to an effector cell such as a muscle. In most of these preparations there is no electrical continuity between cells. Transmission only takes place in one direction, always occurs with a distinct lag and is vulnerable to agents that have no effect on nervous conduction. A much studied preparation of this type is the neuromuscular junction (see figure 7, plate II). From the work of Loewi, Dale, Feldberg, Vogt and others it was known that the chemical released at this synapse is acetylcholine; but it was Katz and his colleagues working in London in the 1950's who provided the basis of our present understanding both of the mechanism of release of the transmitter substance and of the means whereby the liberated acetylcholine elicits a response in the muscle.

The Release of Transmitter

How does the arrival of an impulse at a nerve terminal promote the release of a chemical into the external medium? The simplest answer would be that the impulse alters the permeability of the nerve membrane allowing the transmitter molecules to leak out. This may indeed happen to some extent with acetylcholine which is a small charged molecule, but many transmitters and hormones are peptides and proteins and it is difficult to imagine permeability channels through which these large molecules could escape specifically without loss of other low molecular weight constituents of the cytosol. The key observation was made by Fatt and Katz of University College London who showed that the membrane potential measured by a microelectrode inserted in a muscle close to the termination of a motor nerve is not stable but shows small fluctuations. They noticed that superimposed on a resting potential of almost 100 mV, there were small depolarizations of about 0.5 mV amplitude and characteristic shape (see figure 8, plate III). These seemed to result from the release of acetylcholine because their amplitude was increased by agents that interfere with the breakdown of acetylcholine and they were abolished by curare which blocks the action of acetylcholine on muscle. The simplest hypothesis was that these miniature potentials represented the release of individual molecules of acetylcholine. This hypothesis was easily tested. By applying acetylcholine locally to the surface of the muscle it was soon found that in order to match a miniature potential, not one but about 10 000 molecules of acetylcholine were needed. There seemed no other explanation for this finding than to postulate that the release of acetylcholine is quantized. For instance, acetylcholine may be packaged in some way inside the nerve terminal and a miniature potential represents the release of the contents of one such package. This suggestion, based entirely on electrophysiological observations, has since taken on an almost universal significance in secretory processes because examination in the electron microscope of presynaptic endings and other secretory systems has revealed the presence of small vesicular structures the contents of which seem to be released during secretion (figure 9, plate IV). The precise chain of events seem to involve a process by which the vesicle membrane fuses with the surface membrane of the cell allowing the contents of the intracellular vesicle to diffuse into the extracellular fluid. Subsequently the vesicle membrane is pinched off and returned to the interior of the cell. In a resting nerve, the release of the vesicle contents occurs at a slow random rate which is greatly increased by the arrival of a suitable stimulus.

What is the trigger for vesicle release? Again the work of Katz, Miledi and their colleagues at University College London has provided many of the answers. Transmitter release is a calcium-dependent process and stimulation in the absence of extracellular calcium completely abolishes the release that is normally associated with a nerve impulse. The most direct evidence that the calcium is required inside the cell has come from experiments on a giant synapse in the squid (figure 10, plate V) where Miledi has shown that it is possible to inject calcium into the pre-synaptic terminal and even in the absence of external calcium internal application of calcium promotes transmitter release (figure 11, plate VI). Using the squid giant axon as a model system, Baker and his colleagues have shown that the concentration of ionized calcium in axoplasm is maintained at a very low level—about 0.1 μM—despite the presence of a calcium concentration of about 10 000 μM in the external medium. The arrival of a nerve impulse makes the membrane more permeable to calcium ions. This change in permeability to calcium is brought about by depolarization of the axon membrane but the channel through which calcium enters seems to be quite separate from the sodium and potassium channels that underlie nervous conduction. For the period during which the calcium channels are open, calcium ions rush into the axon raising the intracellular concentration of ionized calcium. As this is initially extremely low, a very small net entry serves to produce a relatively large rise in concentration. At the nerve terminal it seems to be this change in ionized calcium that provides the trigger for initiating transmitter release although the precise mechanism by which calcium initiates secretion is still unknown.

The Post-synaptic Response

Microelectrodes have also been instrumental in discovering how chemical transmitter substances bring about a post-synaptic response. Fatt and Katz and their colleagues showed that stimulation of nerve produces a fall in the resting potential of the muscle, apparently through the near synchronous release of many quanta of transmitter. Provided it is large enough, this reduction in potential sets up an action potential which will propagate along the surface of the muscle in a way essentially similar to that already described in nerves. The post-synaptic response is terminated by either breakdown or removal of the transmitter substance.

The crucial question is: How does a chemical, in this case acetylcholine, bring about depolarization of the muscle membrane? By using two microelectrodes, one in the muscle to record membrane potential and the second as a micropipette for applying acetylcholine to local areas of the muscle membrane, it was shown that in a normal innervated muscle a response to acetylcholine is only found close to the termination of the motor nerve and, even here, acetylcholine is only effective if applied to the outer face of the muscle membrane. The response to acetylcholine is blocked by the South American arrow poison curare, which also blocks transmission, although it is without effect on conduction. All this suggests that specific receptor sites for acetylcholine must be located in the muscle membrane either at or close to the site of the nerve terminal, and the next problem was to determine how reaction at these receptor sites produce the observed changes in potential.

Two kinds of experiment are particularly relevant. The electrical resistance of the muscle membrane is reduced by acetylcholine suggesting that the transmitter increases the number of conducting channels in the membrane, and, irrespective of the initial resting potential of the muscle, acetylcholine always tends to generate a potential change to a level close to zero. This kind of behaviour is to be expected if, in response to acetylcholine, the muscle membrane becomes permeable to *all* small cations, such that its selectivity properties are overridden by the generation of a large number of non-selective cationic channels. This was shown by A. and N. Takeuchi in Japan; their interpretation has been amply confirmed by a variety of other experimental approaches and there is no doubt that the answer to the question "How does acetylcholine depolarize the muscle?" is that it does so by altering the permeability properties of the muscle cell membrane: in the specific case of acetylcholine at the neuromuscular junction by causing a virtually non-selective increase in cation permeability.

To what extent is the neuromuscular junction a good model for all the other synapses both in the periphery of the body and in the central nervous system? Earlier this century Sherrington had shown that many aspects of central nervous function can be described in terms of a balance between what he termed 'central excitatory state' and 'central inhibitory state'. When Eccles and his co-workers re-examined this problem using intracellular microelectrodes they found that at the cellular level the corollary of 'central excitatory state' was depolarization of central neurones and of 'central inhibitory state' was hyperpolarization. It soon became clear that the many lines of communication between neurones in the central nervous system are maintained by a whole battery of chemicals—one set of neurones having the machinery for synthesizing and packaging a particular chemical and another set the specific receptors capable of responding to it. Although the process by which these various chemicals are packaged and released are essentially similar, i.e. they are packaged into vesicles whose contents are released by exocytosis in a calcium-dependent manner, the interaction between chemicals and receptors can give rise to a whole range of different alterations in membrane permeability and longer lasting changes in neuronal metabolism. Those substances whose action results in a fall in membrane potential are excitatory and those substances that hyperpolarize the membrane, for instance by making it more permeable to potassium or chloride, are inhibitory. Whether a particular neurone is caused to fire off an impulse depends entirely on the relative amounts of excitatory and inhibitory inputs that it is receiving at the time.

This transition from the neuromuscular junction of the frog to the mammalian central nervous system has opened up immense vistas, both theoretical and practical, because with the possibility of mapping the chemical wiring diagram of the brain comes the knowledge that by augmenting or blocking specific pathways it may become possible to control many aspects of central nervous function and behaviour.

Already the symptoms of many patients suffering from Parkinson's disease can be alleviated by giving L-dopa which seems to augment the meagre supplies of the transmitter dopamine in their brains and with the recent discovery by Hughes and Kosterlitz in Aberdeen of a whole new class of centrally acting peptides, the enkephalins, that seem to be involved in analgesia there is little doubt that during the next 25 years this area will be one of great activity which one hopes will lead to many drugs of lasting value in the control of mental illness and other disorders of the central nervous system.

Whilst the range and variety of receptors is being explored, microelectrodes are being used to find out more about the ways in which an individual receptor responds to a chemical to produce a change in permeability. Although the ultimate goal of this work is the isolation and purification of receptors, much can be learned about their behaviour from a study of the electrical 'noise' associated with the response of a population of receptors to locally applied transmitters (figure 12, plate VII). Katz and Miledi have shown that information can be obtained about how long a transmitter-induced ion channel remains open and how many ions pass through it. By examining the effects of substances that mimic or compete with a particular transmitter much useful information has come to light about the interaction between drug and receptor at the molecular level. Very recently, by looking at a minute patch of the membrane of a denervated muscle Neher and Sakmann have been able to record the behaviour of a single channel directly thus avoiding the necessity of extracting the information from an analysis of electrical noise. Such an analysis has only been made for a very few transmitters, but for acetylcholine at the frog neuromuscular junction the picture that emerges is one of a channel that opens in response to acetylcholine and remains open depending on temperature for one or a few millisfcconds during which it allows more than 10 000 ions to pass.

In the last few years substances have been found that bind very strongly to the acetylcholine receptor and it has proved possible to obtain receptors in a fairly pure state. So far no detailed chemical structures have been obtained but the isolated receptors have been put to a number of interesting uses. If injected into an animal, the receptor molecules stimulate the production of antibodies which soon start to attack the animal's own acetylcholine receptors. The symptoms generated resemble so closely the human neuromuscular condition known as myasthenia gravis as to suggest that this disease may arise through the body mounting an immunological attack on its own acetylcholine receptors. The extension of this concept to other receptors may have wide implications in medicine.

NEW KINDS OF ELECTRODE

In view of the central importance of ions in the generation of potentials and so many other aspects of cellular function, it would be convenient to have available methods for measuring the activity of particular ionic species inside living cells. Such techniques are gradually becoming available and have already proved of great value in cell physiology. The principle on which most are based is that of the well-known glass electrode for measuring pH. By suitable choice of glass it is possible to make conventional electrodes selective for H^+, Na^+ and K^+ and miniature versions of these electrodes suitable for insertion into cells have now been made by a number of workers. One rather successful design, developed by R.C. Thomas at Bristol, is shown in figure 13, plate VIII. It consists essentially of one microelectrode inside another. The ion-selective glass is drawn into a fine tapering capillary and the narrow end closed. This unit is then sealed inside the tip of a conventional microelectrode which serves to carry the ion-selective glass into the cell. Once inside a cell the potential measured by the electrode is the sum of the cell's resting potential and the potential due to the activity of the ion being measured. The ion activity is obtained by difference between the potential measured by the ion-selective microelectrode and a second conventional microelectrode inserted into the same cell (see figure 14, plate VIII).

Ion selectivity can also be conferred on a microelectrode by filling its tip with an ion exchange resin of suitable ionic selectivity. Electrodes of this type have been used successfully to measure intracellular chloride and are being developed for the measurement of calcium. The monitoring of intracellular calcium by means of an electrode poses acute problems as ionized calcium is present in very low concentrations inside cells and a successful electrode must have very high selectivity for calcium over other intracellular ions: for instance the ratio $[K^+]/[Ca^{2+}]$ inside most cells is $> 10^6$. There is every reason to expect that these technical problems can be overcome and before long a whole range of ion-selective microelectrodes should be available to physiologists making possible a great extension in the range and versatility of electrophysiological measurement.

MOLECULES OF LIFE

D.C. Phillips, F.R.S.[a], Sir Arnold Burgen, F.R.S.[b], J. Feeney[b],
G.C.K. Roberts[b], E.P. Abraham, F.R.S.[c] and S.G. Waley[c]

[a]Laboratory of Molecular Biophysics, Department of Zoology, University of Oxford
[b]Division of Molecular Pharmacology, National Institute for Medical Research, London
[c]Sir William Dunn School of Pathology, University of Oxford

This Chapter consists of three parallel articles:
   Molecular Biology 1952–77, by D.C. Phillips F.R.S.
   Antimetabolites—Drug Action at the Molecular Level, by Sir Arnold
      Burgen, F.R.S., J. Feeney and G.C.K. Roberts
   The Development of the Cephalosporin Family of Antibiotics,
      by E.P. Abraham F.R.S. and S.G. Waley

MOLECULAR BIOLOGY 1952–1977

Molecular Biology is concerned with the structure and function of the very
large molecules upon which life depends, especially the nucleic acids and
proteins, and its explosive growth may be said to have started in 1952 when
agreement was at last reached on the chemical structures and biological roles of
these substances. In that year, reviewing work in their own and other
laboratories, Brown and Todd proposed general chemical structures for
deoxyribonucleic acids (DNA) and ribonucleic acids (RNA) that are shown in
figure 1. In putting forward these structures they noted that "There can be no
question of finality about any nucleic acid structure at the present time since it is
clear that there is no available method for determining the nucleotide sequence".
Nevertheless these general structures for DNA and RNA were soon recognized
as the correct ones. At the same time, following the work of Avery, McLeod and
McCarty on bacterial transformation by DNA which had been published in
1944, it was widely, though by no means universally, believed that the
chemically relatively-inert DNA was the primary store of genetic information in
living organisms—the substance of the genes. This view was advanced further in
1952 by the experiments of Hershey and Chase—an early example, incidentally,
of the importance of radioactive isotopes in experimental molecular

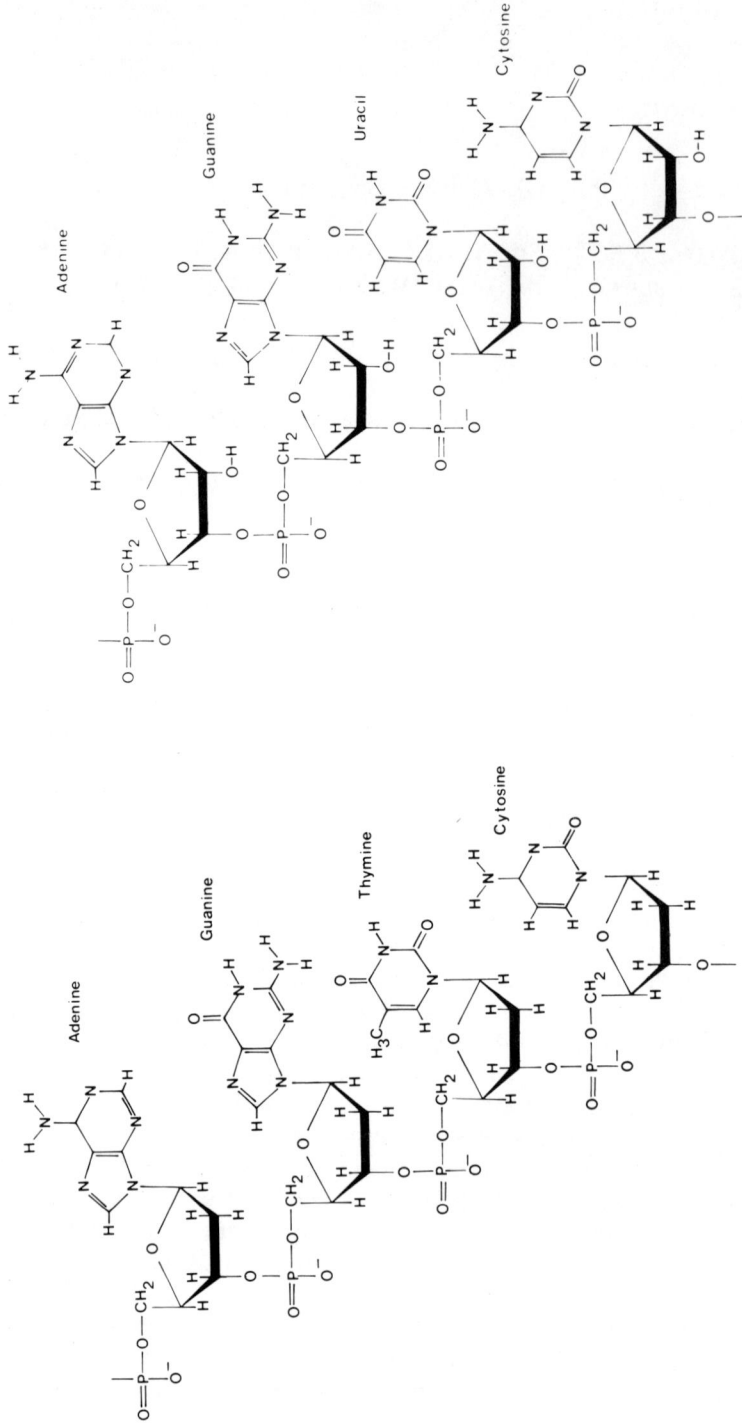

Figure 1. The chemical structure of: left, deoxyribonucleic acid (DNA); right, ribonucleic acid (RNA). The four bases found repeatedly in each type of molecule are shown once each in an arbitrary order.

biology—and there was renewed discussion of how the genes might act as blue-prints for protein synthesis, a process in which the more reactive RNA was known, from the work of Brachet and Caspersson during the 1930's, to be closely concerned.

In 1952 proteins had long been regarded as the most important of all biological substances but their molecular nature was still not established. In a review published in that year Sanger wrote: "As an initial working hypothesis it will be assumed that the peptide theory is valid, in other words, that a protein molecule is built up only of chains of α-amino (and α-imino) acids bound together by peptide bonds between their α-amino and α-carboxyl groups (figure 2). While this peptide theory is almost certainly valid, it should be remembered that it is still a hypothesis and has not been definitely proved. Probably the best evidence in support of it is that since its enunciation in 1902 no facts have been found to contradict it." Sanger's own work on the chemical structure of insulin was soon to lead to the general acceptance of the peptide theory of protein structure and to the recognition that each kind of protein molecule has a unique chemical structure defined by its amino-acid sequence, that is the order in which amino-acid residues of twenty different kinds are joined together in its polypeptide chain or chains.

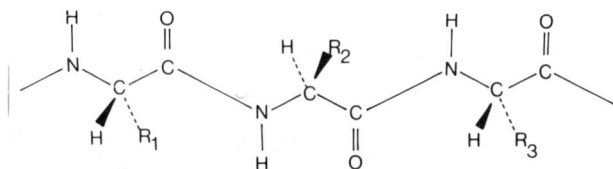

FIGURE 2.   The chemical structure of proteins. Atoms originating in a single α-amino acid are shown grouped together. They are joined by peptide bonds to form polypeptide chains. The symbols R_1, R_2 etc. each represent one of the 20 different side-chains characteristic of the individual amino acids.

Meanwhile discussion of the three-dimensional structures of proteins was already dominated by the peptide hypothesis and by the α-helix and β-pleated sheet structures which had been described by Pauling and Corey in 1951. Almost all of the speakers at a Royal Society Discussion Meeting, held on 1 May 1952 and led by W.T. Astbury, discussed these structures and Edsall, in summing up the discussion, asked: "Is it reasonable to hope that the endless variety of proteins found in nature, and their extraordinarily diverse and specific interactions with one another and with other substances, can be explained on the basis of a few relatively rigid general patterns, simply by varying the nature and sequence of the side-chains attached to the fundamental repeating pattern?"

There were no easy answers to such questions in 1952 but at least the important questions were at last being defined and, in retrospect, it is possible to see what they were. Most importantly perhaps:
(a) how is the information embedded in the DNA of the genes copied for transmission to successive generations?

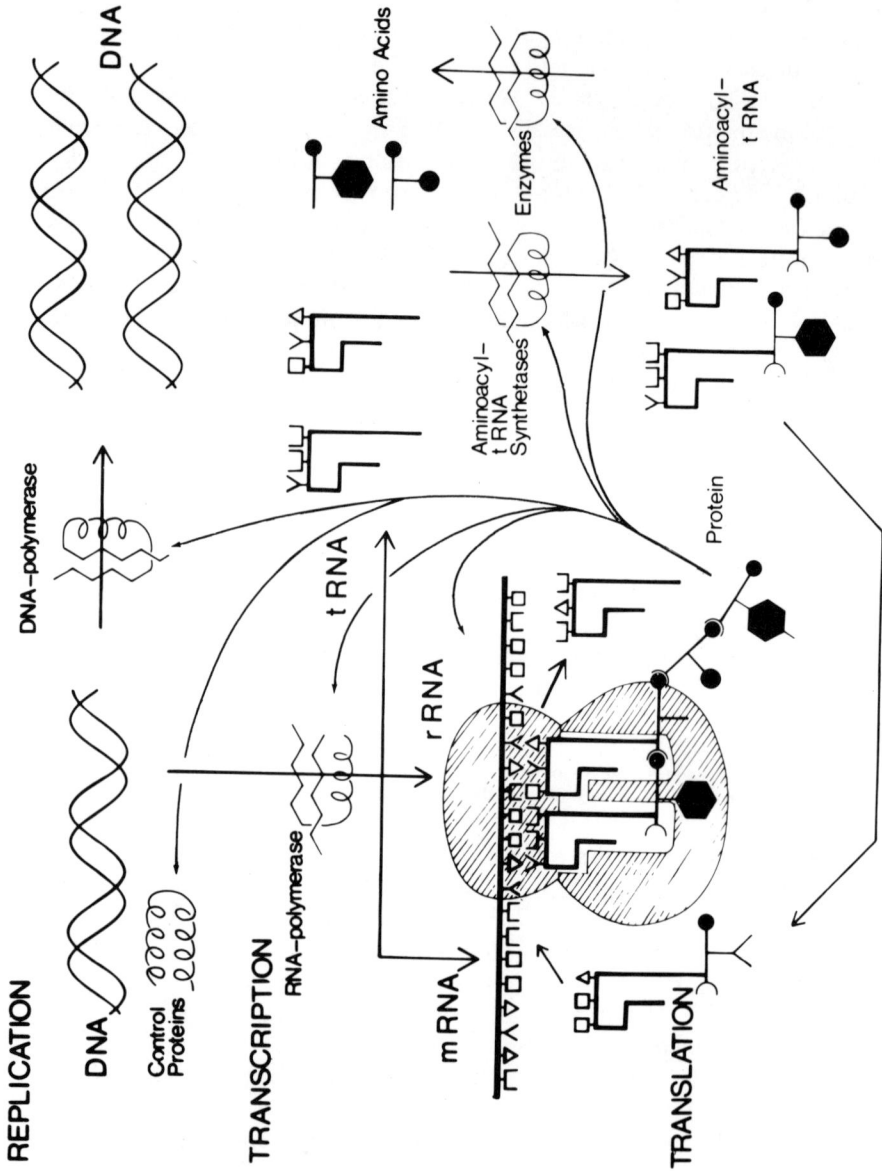

Figure 3.   Schematic diagram showing the replication and transcription of the genetic DNA and its expression in protein synthesis. For description see text.

(b) how does this information direct the synthesis of proteins?

(c) how do proteins, having essentially simple chemical structures, acquire such diverse three-dimensional structures and subtle chemical properties?

The following twenty-five years have provided the answers to these questions that are summarized schematically in figure 3. Here we see, in the briefest summary, that the genetic material (DNA) is supposed to exist in a double helical form, as proposed by Watson and Crick in 1953, in which one strand is complementary to the other: the bases adenine and guanine in one strand are always hydrogen bonded respectively to thymine and cytosine in the other. This complementary structure itself suggested to Watson and Crick the way in principle in which genes are copied for transmission to successive generations of cells or organisms—but it must be emphasized at once that many enzymes, including especially DNA-polymerases must be involved. The synthesis of enzymes, and of all other proteins, requires the expression of the genetic information stored in the DNA and this involves the second general type of nucleic acid, RNA, in which the 2-deoxy-D-ribose sugar of DNA is replaced by D-ribose (figure 1) and the base thymine is replaced by uracil, which can also participate in base pairing with adenine. RNA is transcribed from DNA in the cell nucleus, under the influence of the enzyme RNA polymerase and passes into the cell cytoplasm where it performs three functions in protein synthesis. Three types of RNA are involved: ribosomal RNA (rRNA), messenger RNA (mRNA) and transfer RNA (tRNA). Ribosomal RNA is associated with protein as the major constituent of the ribosomes, the cell organelles of molecular weight about 3×10^6 upon which the protein synthesis takes place. The precise role of ribosomal RNA is not yet understood. mRNA and tRNA, on the other hand, are directly concerned in the flow of information from DNA to protein.

The need for mRNA was discussed by Jacob and Monod in 1961 and experimental evidence for its existence was published in the same year by Brenner, Jacob and Meselson. Its role is to carry the genetic information specifying the structure of a protein from the gene to the ribosomes, the site of protein synthesis. The transfer RNA molecules constitute the adaptors (first proposed by Crick in 1958) that are needed to translate the four-letter language of the genes (the sequences of the four bases A, T, G, C in DNA) into the twenty-letter language of the proteins (the corresponding sequences of the twenty different amino-acid residues that are found in newly synthesized proteins). The genetic code involved is a three-letter code in which three consecutive bases in the mRNA (a codon) determine the nature of the amino acid that is incorporated in the corresponding protein. Through the work of Nirenberg, Khorana and others, which was described in Crick's Croonian lecture to the Royal Society in 1966, this code is completely known (table 1).

The adaptor or tRNA molecules are relatively small RNA molecules each containing between 73 and 93 ribonucleotides in a single chain. There is a different one for each codon and, following the work of Holley, the ribonucleotide sequences of many of them have been established. They are closely similar to one another and presumably to yeast phenylalamine tRNA, the

Table 1

The Genetic Code

The positions of the bases adenine (A), cytosine (C), uracil (U) in the mRNA codons are read from the 5' end of the polynucleotide chain. The amino acids are represented by their standard abbreviations.

5'-OH Terminal base	Middle base				3'-OH Terminal base
	U	C	A	G	
U	UUU ⎱ Phe UUC ⎰ UUA ⎱ Leu UUG ⎰	UCU ⎱ UCC ⎬ Ser UCA ⎪ UCG ⎰	UAU ⎱ Tyr UAC ⎰ UAA Stop UAG Stop	UGU ⎱ Cys UGC ⎰ UGA Stop UGG Trp	U C A G
C	CUU ⎱ CUC ⎬ Leu CUA ⎪ CUG ⎰	CCU ⎱ CCC ⎬ Pro CCA ⎪ CCG ⎰	CAU ⎱ His CAC ⎰ CAA ⎱ Gln CAG ⎰	CGU ⎱ CGC ⎬ Arg CGA ⎪ CGG ⎰	U C A G
A	AUU ⎱ AUC ⎬ Ile AUA ⎪ AUG Met & Start	ACU ⎱ ACC ⎬ Thr ACA ⎪ ACG ⎰	AAU ⎱ Asn AAC ⎰ AAA ⎱ Lys AAG ⎰	AGU ⎱ Ser AGC ⎰ AGA ⎱ Arg AGG ⎰	U C A G
G	GUU ⎱ GUC ⎬ Val GUA ⎪ GUG ⎰	GCU ⎱ GCC ⎬ Ala GCA ⎪ GCG ⎰	GAU ⎱ Asp GAC ⎰ GAA ⎱ Glu GAG ⎰	GGU ⎱ GGC ⎬ Gly GGA ⎪ GGG ⎰	U C A G

three-dimensional structure of which has been determined independently by Klug and Rich and their colleagues using X-ray analysis. Each has a chemical structure in which extensive base pairing is possible so that the course of the polyribonucleotide chain is well represented by the outline of a clover leaf though, in three dimensions, the molecules are found to be L-shaped with three bases exposed at the end of one arm and the site of attachment for a particular amino acid at the end of the other arm. The three exposed bases form the anti-codon, that is they are complementary, in terms of the base pairing found in double-helical DNA, to the codon for the particular amino acid that is carried at the end of the other arm of the molecule. The association of the tRNA molecules with their corresponding amino acids is brought about by enzymes, the amino-acyl-tRNA synthetases (at least one of which exists for each amino acid).

Protein synthesis proceeds residue-by-residue on the ribosome as it moves in relation to the mRNA. Successive tRNA molecules, corresponding to the codon triplets, deliver their amino acids to the site of synthesis and these are incorporated in the growing polypeptide chain which is extruded from the ribosome, amino end first. Again many enzymes are involved.

One important aspect of protein synthesis remains to be mentioned. The sequence of bases in the DNA of the gene determines the sequence of amino-acid residues in the corresponding protein in the way that has been outlined but no additional information is available to determine the three-dimensional structure of the protein. The three-dimensional structure adopted by the protein depends only upon its amino-acid sequence and the environment in which it finds itself. The relationship between the amino-acid sequences and three-dimensional structures of proteins has therefore been an intensive field of study, made possible by the crystallographic analysis of protein structures.

Globular proteins, especially enzymes, have been crystallized in increasing numbers since Sumner's first success with urease and attempts to determine their structures in detail by X-ray diffraction began in 1934 with the study of pepsin by Bernal and Crowfoot. Despite intensive efforts by many workers, however, success hardly seemed possible until 1954 when Green, Ingram and Perutz showed that the method of isomorphous replacement could be used in protein structure analysis. Since then the structures of more than fifty globular proteins have been determined in atomic detail by this method, starting with Kendrew's study of myoglobin, an oxygen storage protein, in 1958–60. The structure of lysozyme, the first enzyme to be described in this detail, was published in 1965. Crystallographic and other experiments have also shown how many of these molecules perform their biological functions so that, for example, the complex behaviour of haemoglobin that is involved in the transport of oxygen from the lungs to other tissues, and the catalytic activity of lysozyme and other enzymes can be understood in atomic detail.

Most strikingly nearly all of these protein structures are built up from α-helices and β-strands to an extent that Edsall hardly dared hope when these elements of secondary structure were first discussed. This finding has had two main consequences. First, intensive efforts have been made to establish firm rules relating these structures to the underlying amino-acid sequences, with more than enough promise of success to maintain a high level of activity in this field. Secondly, there has been much discussion of the resemblances between different protein structures especially in relation to their evolution: do similarities arise because particular conformations are energetically favoured (or more easily achieved during folding) or do they reflect a common descent from the same ancestral protein? These different models of convergent and divergent evolution each have their adherents but the question perhaps is really part of the more fundamental question: how did the complex apparatus and mechanisms of protein synthesis evolve, dependent as they are upon the intimate interplay of both nucleic acids and proteins?

In principle it may never be possible to answer this question in detail but current work is providing further details of the molecular mechanisms that are involved in gene replication and expression and, together with some dramatic surprises, new evidence about possible evolutionary processes is appearing.

Structural studies by diffraction methods, involving further development of X-ray crystallography and electron microscopy, associated with chemical studies are revealing the structures and properties of increasingly complex molecules and molecular systems. Only a few examples can be quoted but clearly studies on aminoacyl-tRNA synthetases, the enzymes responsible for attaching the activated amino acids to their appropriate tRNA molecules, are of fundamental importance since these enzymes play an essential part in maintaining the fidelity of gene translation. Crystallographic analyses are well advanced but chemical and kinetic studies have already shown in principle how these enzymes attain such a high degree of specificity. They appear to have two related active sites in one of which formation of the correct amino-acid-tRNA linkage is favoured while incorrectly paired linkages are destroyed in the other. This editing mechanism enables the enzymes to discriminate between very similar amino acids: for example, isoleucine and valine differ only by one methyl group yet the frequency with which valine is incorporated in proteins in place of isoleucine is barely one in 3000. Many of the other proteins involved in replication, transcription and translation are also being studied, especially the proteins of the ribosomes. These organelles, which are usually supported on membranes within the cell, are very large and complex aggregates of protein and RNA and, although progress is being made by Wittmann and others in analysing their structures chemically, and by diffraction methods, few details are yet clear. Similarly promising results have been obtained in studies of the organization of DNA within the chromosomes, where it is complexed with protein to give a highly condensed conformation. But the most detailed results on the important topic of protein–nucleic acid interactions at the moment are coming from studies of tobacco mosaic virus (TMV) that were begun by Bernal and Fankuchen in 1941. TMV is a rod-like virus, 3000 Å long, in which identical protein subunits are arranged in a helix of 49 subunits in three turns protecting the RNA. Using novel methods to analyse the fibre diffraction pattern given by gels of these particles, Holmes and his colleagues have produced an electron-density map in which the general fold of the protein chain is clear and the conformation of the RNA and its interaction with the protein can be seen. In parallel with this work, Klug and his associates have studied single crystals of TMV protein in which the subunits are arranged in 17-fold rings. Comparison of the two sets of results is revealing the virus structure in detail and showing directly for the first time extensive protein-nucleic acid interactions. Evidence on virus assembly is also emerging. Just as protein chains fold spontaneously to their active conformations in the appropriate environment, so do larger aggregates, such as ribosomes and viruses, assemble themselves from their individual protein and nucleic acid components. These studies of TMV are showing how this is achieved through intermediate disc structures related to that being studied in the crystals.

But the most dramatic and surprising results recently have come from chemical studies. Providing the evidence called for by Brown and Todd in 1952, Fiers in 1975 determined the complete ribonucleotide sequence of the RNA in bacteriophage MS2 and Sanger has now determined the complete sequence of nucleotides in the DNA of bacteriophage ΦX174. This latter sequence comprises some 5375 nucleotides and, like the earlier work on MS2, it shows in detail the coding for the various proteins that are known to be synthesized according to the instructions embedded in this genetic material including, of course, the nucleotide sequences that signal the beginnings and ends of the protein chains. In addition, however, the deoxyribonucleotide sequence of ΦX174 shows that the individual proteins are not all coded separately in the DNA. At least two pairs of genes overlap so that the same nucleotide sequences are read (in triplets) from different starting points to code for two completely different amino-acid sequences (table 2).

TABLE 2

Part of Nucleotide Sequence of Bacteriophage ΦX174 DNA
showing simultaneous coding for two proteins, A and B

The stop signal (TGA) for protein B is included. Since this is a DNA sequence, T (for Thymine) replaces the U (for Uracil) shown in the diagram of the genetic code (figure 4).

Protein A	Ser	Phe	Ile	Ala	Ser	Met
DNA	G A G T T T T A T C G C T T C C A T G A					
Protein B	Glu	Phe	Tyr	Arg	Phe	His	Asp

Protein A	Thr	Gln	Lys	Leu	Thr	Leu	Ser
DNA	C G C A G A A G T T A A C A C T T T C G						
Protein B		Ala	Glu	Val	Asn	Thr	Phe

Protein A	Asp	Ile	Ser	Asp	Glu	Ser
DNA	G A T A T T T C T G A T G A G T C G A A					
Protein B	Gly	Tyr	Phe	*Stop* B		

Even more amazingly, Shaw and his colleagues have discovered very recently that the coding capacity of the DNA in bacteriophage G4 is used in all three reading frames. Thus in one TGATG sequence, A is recognized as part of the

ATG 'start' codon of one gene, as part of the TGA 'stop' codon of a second gene and as part of the GAT codon for aspartic acid in a third gene (cf. table 1). Clearly DNA has much more coding capacity than had been supposed possible on the assumption that each gene is physically separate.

Similarly detailed studies are now beginning to be made of the genes of higher organisms and again surprising results are appearing. Here the remarkable finding is that the nucleotide sequences in the DNA that are translated into amino-acid sequences by the processes described above are not found as continuous sequences in the chromosomes but are interrupted by lengths of DNA that are not translated. Thus, for example, Jeffreys and Flavell have shown that the gene coding for the β-chain of rabbit haemoglobin is interrupted by at least one piece of DNA some 700 nucleotides long that does not correspond to any part of the globin sequence or appear in the equivalent mRNA. Apparently a gene in higher cells is a mosaic in which expressed sequences are held, as Gilbert has put it, in a matrix of 'silent' DNA. These remarkable findings help to explain a long-standing puzzle, that much of the RNA produced in cell nuclei is not transmitted to the cytoplasm but is degraded within the nuclei as Harris showed in 1959. Presumably all the DNA that embraces the genetic information coding for a protein is transcribed to RNA and the part corresponding to the silent DNA is excised and degraded during subsequent preparation of the mRNA.

Like the earlier observation of rapidly-turning-over nuclear RNA, these new findings suggest mechanisms that may play a part in evolution and may be found to have a bearing on the resemblances that have been observed between different proteins. Hitherto it has been supposed that single base changes in the DNA of the genes, brought about by mutagenic agents of various kinds, change one amino-acid residue in the corresponding protein. Now it appears that such changes occurring in the regions that determine whether the DNA is transcribed to mRNA, or only to RNA that is excised and degraded in the nucleus, might have much more dramatic effects. For example, they might bring about the incorporation of extended sequences of amino-acid residues corresponding to the 'silent' DNA or to more distant DNA sequences or the deletion of extended sequences no longer represented in the mRNA.

These and many other results show that research in Molecular Biology is still dramatically exciting and that the scheme of figure 3 may, after another twenty-five years, seem a very primitive picture. Nevertheless, it represents a remarkable achievement that underlies all modern thinking in biology and has already had a profound effect upon medical research and practice.

ANTIMETABOLITES—DRUG ACTION AT THE MOLECULAR LEVEL

A significant part of the success of modern medicine depends upon the availability of drugs which have highly specific effects, either on some aspect of bodily function or on invasive cells, such as bacteria, parasites or indeed tumour cells. Many, perhaps most, of these drugs produce their effects by acting as antimetabolites. An antimetabolite is a molecule which competes with an endogenous compound for binding to a site in the cell, thus disturbing the function of the cell. The site may be an enzyme, a cell-surface receptor for a hormone or neurotransmitter, or a nucleic acid. As our knowledge of cellular metabolism has increased, it has become increasingly possible to design compounds to interfere with particular metabolic processes.

The Origin of the Idea of Antimetabolites

The first synthetic compounds to be recognized as antimetabolites arose from work based on a rather different rationale. In the early years of this century, Ehrlich showed that dyes could be effective therapeutic agents. Many such compounds were tested, and in 1935 Domagk found that Prontosil (figure 4) had considerable antibacterial activity. However, it was soon shown by Trefouel that Prontosil is metabolized to sulphanilamide (figure 5) and that it is the latter compound which is responsible for the antibacterial action. In 1940, Woods showed that the effects of sulphanilamide were reversed by p-aminobenzoic acid (figure 5). Noting the structural resemblance between the two molecules, he

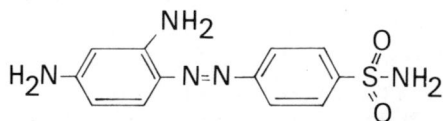

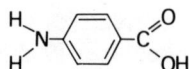

FIGURE 4.   The structure of Prontosil.

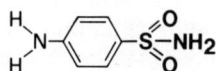

Sulphanilamide   p–Aminobenzoic Acid

FIGURE 5.   The structure of sulphanilamide, the active metabolite of Prontosil, compared with that of p-aminobenzoic acid, the natural compound with which it competes.

suggested that sulphanilamide acts as an antimetabolite of p-aminobenzoic acid. It was later found that p-aminobenzoic acid is one of the precursors of the folate coenzymes. Sulphanilamide acts by inhibiting the enzyme (dihydropteroate synthetase) for which p-aminobenzoic acid is a substrate (figure 6).

The antimetabolite theory was quickly recognized as an attractive approach to the design of new drugs and when the structure of the vitamin folate (figure 7)

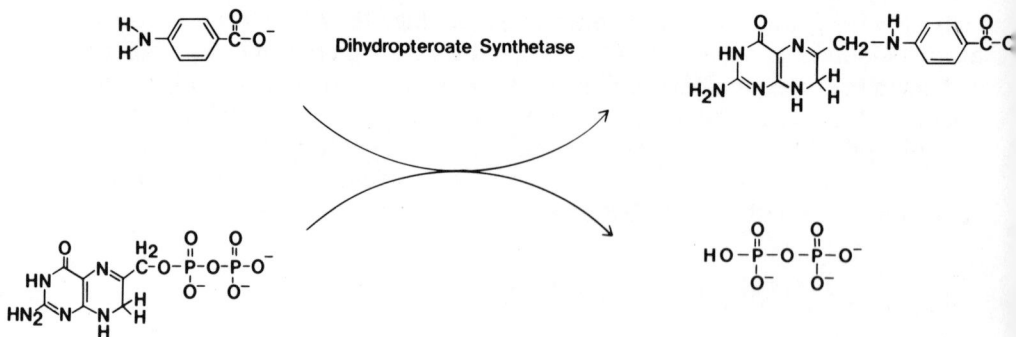

FIGURE 6. The reaction catalysed by dihydropteroate synthetase. p-Aminobenzoic acid (top left) is incorporated into dihydropteroic acid (top right), replacing the pyrophosphate group of the precursor 6-hydroxymethyl-pterin pyrophosphate (bottom left).

was established in 1946 attempts were soon made to synthesize antimetabolites to it. The first two such compounds were aminopterin and methotrexate (figure 7) which were soon shown by Farber and others to be effective in the treatment of leukaemia. Since then, many 'antifolate' drugs have been synthesized, and several have found clinical application. Methotrexate, often in combination with other drugs, is still used with success in the treatment of certain forms of cancer, the best results being obtained in acute lymphocytic leukaemia, choriocarcinoma and Burkitt's lymphoma. Pyrimethamine and cycloguanil are used in the treatment of malaria, while trimethoprim in combination with a sulphonamide is a most effective antibacterial agent. All these 'antifolate' drugs act by inhibiting the enzyme dihydrofolate reductase (figure 8). Tetrahydrofolate, the product of the action of this enzyme, acts as an essential coenzyme in cellular metabolism, functioning as a 'carrier' of one-carbon fragments. An active dihydrofolate reductase is essential to maintain the levels of tetrahydrofolate derivatives in the cell; if the enzyme is inhibited and these levels fall, the cell becomes unable to synthesize deoxythymidylate (dTMP) or purines, both essential precursors for DNA synthesis.

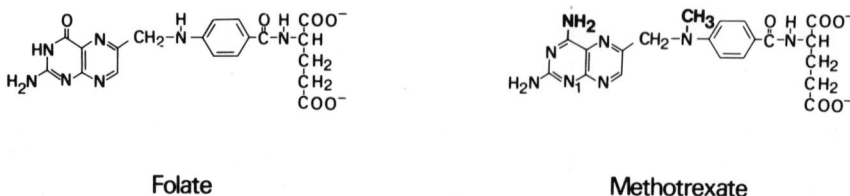

Folate   Methotrexate

FIGURE 7. The structure of the vitamin folate, compared with that of its antimetabolite, methotrexate. The structural differences are indicated in bold type. The related antimetabolite, aminopterin, lacks the methyl group on N-10 and thus resembles folate even more closely.

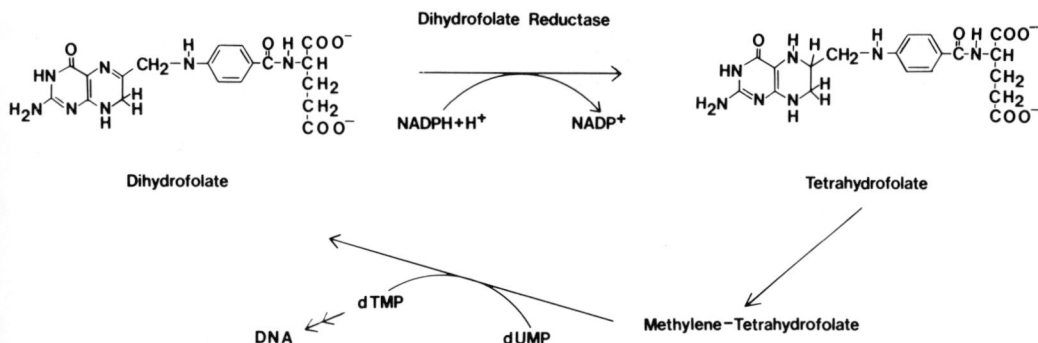

FIGURE 8. The reaction catalysed by dihydrofolate reductase, and its relationship to the synthesis of deoxythymidylate (dTMP) and DNA.

Principles of Antimetabolite Design

The antimetabolite must clearly be sufficiently similar in structure to the natural compound to bind tightly to the relevant site, but sufficiently different to be non-functional—in the case of enzymes, with which we are concerned here, an inhibitor, not a substrate. However, the prerequisite for success in antimetabolite chemotherapy is *selective toxicity*. It is necessary to interfere with the metabolism of the invasive cells without doing too much damage to that of the host. There are several ways in which this selectivity can be achieved. First, one can inhibit a reaction which takes place only in the invasive cells. This is the basis of the usefulness of sulphonamides as antibacterial agents, since folate biosynthesis does not take place in mammals, which rely on dietary folate. Alternatively, it may be possible to design a compound which binds selectively to the enzyme of the invasive cells. As shown by Hitchings and his colleagues, trimethoprim is an effective antibacterial agent because it binds several thousand times more tightly to bacterial than to mammalian dihydrofolate reductase, and a similar species-selective binding underlies the effectiveness of pyrimethamine against malaria. Finally, the consequences to the cell of inhibition of a particular metabolic process may differ from one cell type to another. For example, rapidly dividing cancer cells are more seriously affected by inhibition of DNA synthesis than are normal cells; this provides part of the explanation for the effectiveness of methotrexate against some forms of cancer.

The first stage of rationality in drug design is thus based on 'mimicking' the structure of a natural metabolite. A systematic exploration of structural modifications in search of potency and selectivity is required, and in general a very large number of compounds must be synthesized and screened. It is not uncommon to find that the potency or selectivity of antimetabolites depends upon structural features which have no counterpart in the natural compound. Examples in the sulphonamide and antifolate series are sulphadiazine and trimethoprim (figure 9). Clearly there must be 'ancillary' binding sites on the

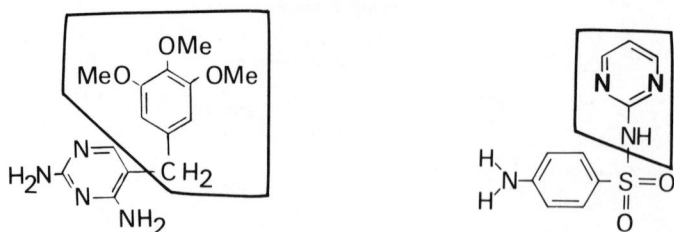

Trimethoprim   **Sulphadiazine**

FIGURE 9.   The structures of sulphadiazine and trimethoprim, two effective antimetabolites—a sulphonamide and an antifolate respectively. The enclosed parts of the structures are essential to the effectiveness of these drugs, yet they have no counterpart in the structures of the natural compounds with which they compete.

enzyme with which these groups interact—sites close to but distinct from the substrate binding site. This opens up new possibilities for the design of selective antimetabolites, such as trimethoprim, but it also means that simple 'mimicking' of the natural compound is no longer a sufficient guideline for synthesis.

The second stage of rationality would be to design drugs to fit a binding site of known structure. With our increasing knowledge of the structure of binding sites on proteins from X-ray crystallography, this is becoming more and more feasible—a recent example is the work of Goodford and his colleagues on the diphosphoglycerate binding site of haemoglobin. Clearly a prerequisite for this approach is a detailed understanding of the way in which the substrate and any known inhibitors interact with the binding site. Apart from knowing the three-dimensional structure of the complex in some considerable detail, we must be able to describe the changes in conformation of the protein and/or the small molecule which occur when they interact with one another. In addition, we must understand the chemical or conformational reasons for the inactivity of the antimetabolites. Sometimes this is obvious—for example trimethoprim simply lacks the ring which is reduced by dihydrofolate reductase—but often more subtle effects are responsible, as in the case of methotrexate, which is not a substrate in spite of its close structural resemblance to folate. Finally, to design compounds which will bind tightly to the known structure, we need to know what is the contribution of each of the protein–small-molecule contacts to the overall energy of interaction; this question has yet not been adequately answered in any system. In attempting to gather together all the necessary information, we must call on all the experimental and theoretical resources of molecular biology. In the remainder of this article we shall describe some of the approaches we have been using in attempting to reach this level of understanding in the case of dihydrofolate reductase.

Dihydrofolate Reductase

As noted above, a number of potent and therapeutically useful inhibitors of dihydrofolate reductase are already known. A convenient starting point is thus to compare the binding of folate (a substrate) with that of the potent inhibitor methotrexate, which binds up to 10 000 times more tightly.

The essential structural difference between folate and methotrexate is the replacement of the 4-oxo group by an amino group. One of the chemical repercussions of this change is a substantial change in the dissociation constant for protonation at N–1 on the pteridine ring; the pK_a increases from 2.0 for folate to 5.35 for methotrexate. Thus methotrexate is more readily protonated than folate, raising the possibility (first noted by Baker) that a difference in ionization state may contribute to the difference in binding. Since the protonated and neutral molecules have quite distinct ultraviolet absorption spectra, we can investigate this point by studying the changes in the ultraviolet spectra of the small molecules when they bind to the enzyme.

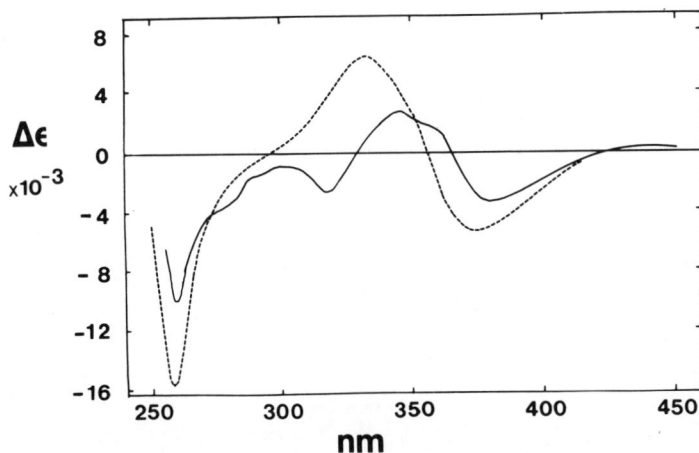

FIGURE 10. The ultraviolet difference spectrum (see text) produced when methotrexate binds to dihydrofolate reductase is shown as the solid line. The dotted line is the difference spectrum between protonated and neutral methotrexate. (From Hood and Roberts, *Biochem. J.* **171**, 357, 1978.)

This is most conveniently done by difference spectroscopy. In this technique two separate sample cells are used, each having two compartments. One cell, the reference cell, contains enzyme solution in one compartment and a solution of the small molecule in the other; the other cell contains exactly the same solutions but the contents of the two compartments are mixed, allowing the small molecule to bind to the enzyme. By recording the difference in light absorption between the two cells, using a double-beam spectrophotometer, we can thus measure only the *changes* in the absorption spectrum which occur on binding. Such a difference spectrum, generated when methotrexate binds to dihydrofolate reductase from *Lactobacillus casei*, is shown as the solid line in figure 10. It is compared with the difference spectrum between protonated and neutral methotrexate (the dotted line). Both spectra have in common the negative bands at 260 nm and 380 nm, and the spectrum generated when methotrexate binds to the enzyme can be shown to consist of a contribution from protonation of the methotrexate on binding, and a contribution (responsible for the features between 320 and 360 nm) from the change in its environment on binding. A quantitative analysis shows that the pK_a of methotrexate increases from 5.35 to 8.55 on binding to dihydrofolate reductase—the protonated form binds 1600 times more tightly than the neutral molecule. In contrast, a similar analysis of the difference spectrum produced when folate binds to the enzyme shows that it binds as the neutral molecule. This is an important difference between the natural compound and its antimetabolite, and accounts for about one-third of the difference in binding energy between the two at pH 7.5.

There is increasing evidence that at least part of the remaining difference in binding energy between folate and methotrexate originates from a conformational difference between the two complexes. Some such evidence comes from attempts to assess the importance of different parts of the methotrexate molecule for binding. One approach to this is the study of 'fragments'; we have used 2,4-diaminopyrimidine (DAP) and N-(p-aminobenzoyl)-L-glutamate (PABG) as 'fragments' of methotrexate (figure 11). As one might expect from their structural relationship to methotrexate,

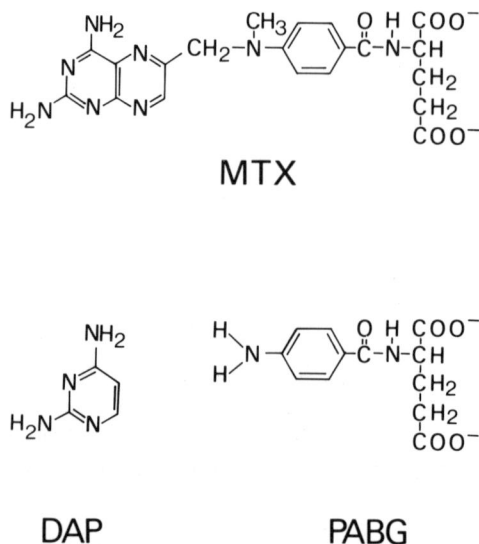

MTX

DAP   PABG

FIGURE 11. The structures of methotrexate (MTX) and its 'fragments' 2,4-diaminopyrimidine (DAP) and N-(p-aminobenzoyl)-L-glutamate (PABG).

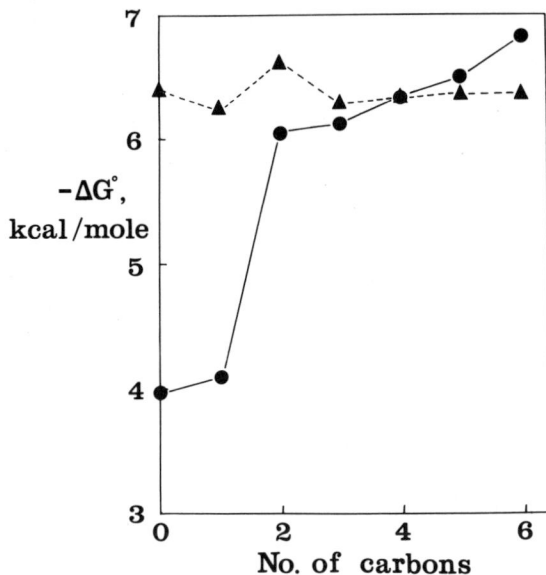

FIGURE 12. The binding of a series of N-(p-alkylaminobenzoyl)-L-glutamates to dihydrofolate reductase, expressed as the change in free energy on binding as a function of the number of carbon atoms in the alkyl chain. The circles and solid line refer to the binding to the enzyme alone, the triangles and dashed line to binding to the enzyme-2,4-diaminopyrimidine complex.

these two molecules can bind simultaneously to the enzyme. However, they also bind co-operatively: PABG binds some 50 times more tightly to the enzyme–DAP complex than to the enzyme alone. This co-operativity appears to arise from a conformational change produced by the binding of the first 'fragment' which increases the affinity of the enzyme for the second 'fragment'. To explore this co-operative binding further, we have studied a series of N-(p-alkylaminobenzoyl)-L-glutamates (see figure 12). Increasing the length of the alkyl chain markedly increases the binding of these compounds to the enzyme, indicating that the alkyl chains are interacting with an ancillary binding site on the enzyme. However, in the presence of DAP, the alkyl chains have essentially no effect on binding. The binding of DAP has produced a conformational change which in some way makes the ancillary binding site inaccessible; it produces a change in specificity as well as a change in affinity.

There is good evidence from nuclear magnetic resonance spectroscopy (see below) that these 'fragments' bind in a way which is closely similar to the mode of binding of the corresponding parts of the whole methotrexate molecule. The conformational changes which accompany 'fragment' binding must also occur when methotrexate binds, so that the binding of one end of the methotrexate molecule will influence the binding of the other end. This clearly has important

DHFR + TRIMETHOPRIM

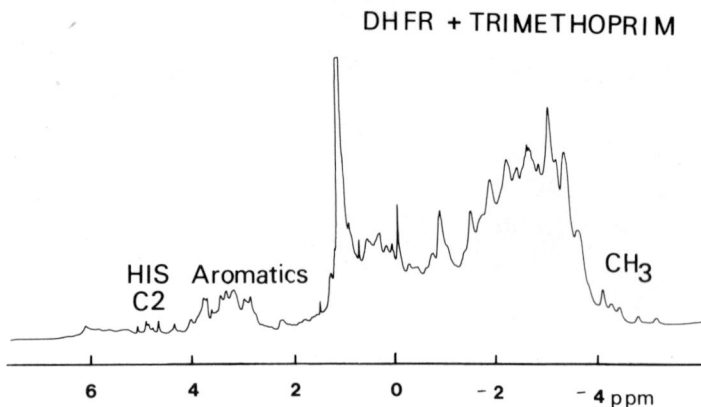

FIGURE 13. The 270 MHz ^{1}H nuclear magnetic resonance spectrum of the dihydrofolate reductase–trimethoprim complex. The sharp line at 0 ppm is the dioxan chemical shift reference, and the large peak at about 1 ppm arises from the residual protons in the ^{2}H$_2$O solvent.

implications for the understanding of structure–activity relationships in this system.

Interactions with Individual Residues: Nuclear Magnetic Resonance Spectroscopy

If we are to define with any precision the nature of the various conformational changes which occur when small molecules bind to dihydrofolate reductase, we must use a method which allows us to examine effects on individual amino-acid residues. In recent years, nuclear magnetic resonance (n.m.r.) spectroscopy has been used increasingly in studies of the binding of small molecules to proteins in solution, since not only does it allow one to look at individual amino-acid residues but at individual nuclei in the protein. Many nuclei—for example, ^{1}H, ^{13}C (but not ^{12}C), ^{19}F, ^{31}P and numerous others—have magnetic moments. If a proton, for example, is placed in an external magnetic field it will occupy one of two states (or energy levels), which can be regarded as corresponding to orientation of the nuclear magnetic moments *with* the field (lower energy) or *against* the field (higher energy). As in other forms of spectroscopy, we can induce transitions from the lower to the upper energy level by supplying the necessary energy in the form of electromagnetic radiation of a frequency appropriate to the difference in energy between the two levels. For nuclear magnetic resonance (n.m.r.) spectroscopy, the appropriate frequency is in the radio-frequency range; it depends on the nucleus being observed, and is directly proportional to the strength of the external magnetic field: for example at a field of 2.35 tesla (23.5 kilogauss) the resonance frequency for ^{1}H will be 100 MHz, for ^{13}C 25.2 MHz and for ^{2}H 15.4 MHz. The chemical and biochemical usefulness of n.m.r. spectroscopy arises from the fact that the precise resonance frequency for a nucleus in a molecule depends in a sensitive way on the chemical environment in which it finds itself. Not only do the protons in a methyl group have a different resonance frequency from those on an aromatic ring, but methyl groups in different environments—such as buried inside a protein or exposed to the solvent, or in a ligand bound to the protein or free in solution—will also show significantly different resonance frequencies. Thus the ^{1}H n.m.r. spectrum of dihydrofolate reductase (figure 13) contains a wealth of information: there is a resonance in the spectrum from each of approximately 1000 protons in the protein, and the precise position of each of these resonances depends upon the environment of the corresponding proton in the protein, and will change on the binding on a small molecule if the environment of the proton changes. Thus very

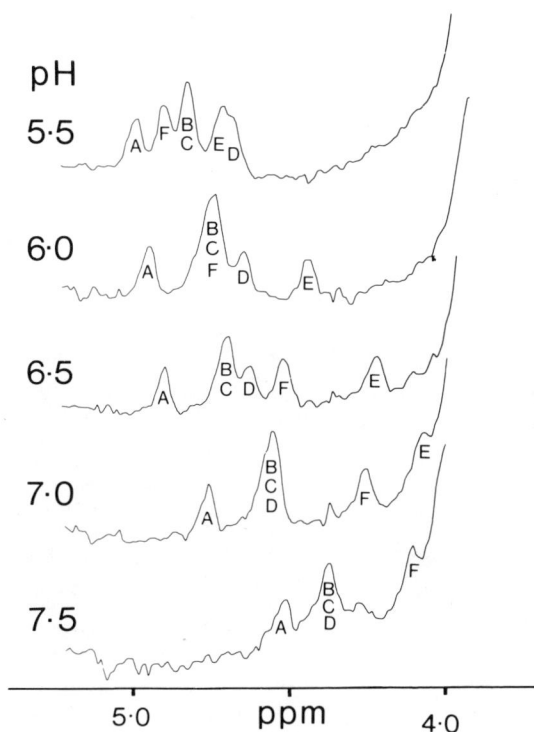

FIGURE 14. The pH dependence of the histidine C2–H resonances of dihydrofolate reductase in its complex with trimethoprim. The individual resonances are labelled A–F. (From *Proc. Roy. Soc. B.*)

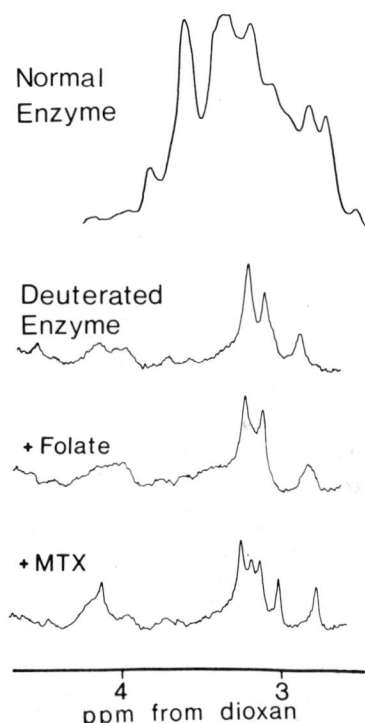

FIGURE 15. The aromatic region of the ^{1}H nuclear magnetic resonance spectra of normal dihydrofolate reductase (top) and of a selectively deuterated dihydrofolate reductase for which the only resonances in this region of the spectrum are those of the five tyrosine residues. The effects of adding folate or methotrexate (MTX) are shown in the lower two spectra. (From Roberts, Feeney, Birdsall, Kimber, Griffiths, King and Burgen, in: *NMR in Biology*, Academic Press 1977.)

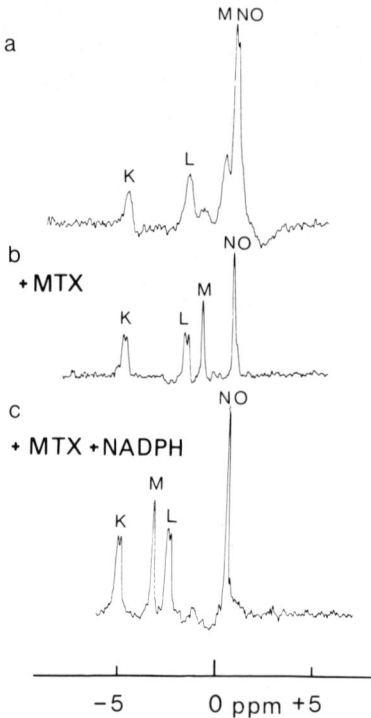

FIGURE 16. The ^{19}F nuclear magnetic resonance spectra of a dihydrofolate reductase containing 6-fluoro-tryptophan in place of the normal tryptophan residues. Spectrum a is that of the enzyme alone, and spectra b and c show the effects of adding methotrexate (MTX) followed by NADPH. The letters K–O identify the resonances of individual 6-fluoro-tryptophan residues. (From Kimber, Griffiths, Birdsall, King, Scudder, Feeney, Roberts and Burgen, *Biochemistry* **16**, 3492 (1977).)

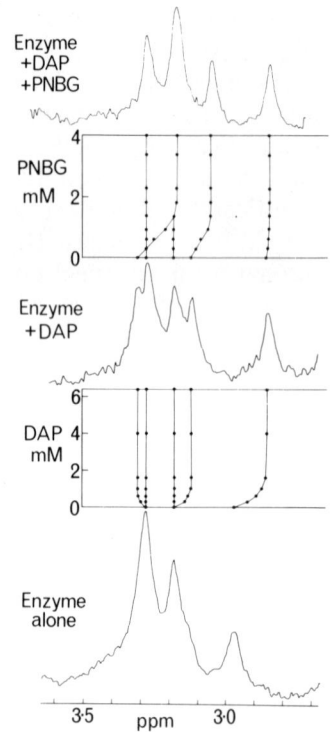

FIGURE 17. The effects of adding increasing concentrations of 2,4-diaminopyrimidine (DAP), followed by p-nitrobenzoyl-L-glutamate (PNBG), on the 2,6-proton resonances of the five tyrosine residues of selectively deuterated dihydrofolate reductase. (From *Proc. Roy. Soc. B.*)

detailed information on the binding process is available, provided that one can resolve resonances of individual amino-acid residues. For example, the positions of the resonances of the C2 protons of the six histidine residues depend markedly on the ionization state of the residue. This is illustrated in figure 14 for the enzyme–trimethoprim complex; we can use this effect to determine the pK_a values of individual residues, and to define the effects of ligand binding on them. Histidine residues H_A and H_F are affected in a very similar manner by the binding of either folate or methotrexate; the pK_a of residue H_F is increased by about 0.7 units, and this may well be an effect of the glutamate carboxylate groups of the ligands. In contrast, inhibitors and substrates have *opposite* effects on the pK_a of histidine H_E folate, producing an increase while methotrexate binding produces a decrease in pK_a.

It is clear from figure 13 that the complexity of the ^{1}H n.m.r. spectrum of dihydrofolate reductase is such that it is very difficult to resolve any more resonances from individual residues. A general solution to this problem is provided by the procedure of selective deuteration. If the protein is isolated from an organism grown on largely deuterated amino-acids, many of the ^{1}H resonances are removed from the spectrum, which is thus greatly simplified. The sort of simplification which can be achieved in this way is illustrated in figure 15, which shows the region of the spectrum containing the resonances of the protons on the aromatic rings of the phenylalanine, tyrosine, tryptophan and histidine residues. The top spectrum is that of the normal protein, while the others are from an analogue in which all the aromatic protons *except* the 2,6-protons of the five tyrosine residues have been replaced by deuterium. We can now examine the effects of ligand binding on individual tyrosine residues; folate binding affects only a single residue, while methotrexate binding affects three. An alternative approach to simplification of the spectrum is to incorporate a fluorine-labelled amino-acid and examine the ^{19}F n.m.r. spectrum; figure 16 shows the ^{19}F spectrum of dihydrofolate reductase containing 6-fluorotryptophan. Again, resonances of individual residues are readily resolved, and we can see that one fluorotryptophan resonance is markedly affected by the binding of methotrexate and coenzyme. Folate binding produces a much smaller shift of this resonance. For many of the residues in the protein, there are clear differences in the effects of methotrexate and folate binding which, at least in part, reflect differences in the conformational changes which these small molecules produce.

We can also use n.m.r. spectroscopy to investigate the conformational changes responsible for the co-operative binding of the 'fragments' of methotrexate (figure 17). If we compare the effects of DAP and PABG binding, alone and together, we note that in many cases their effects are distinctly non-additive. For example, the resonance position of tyrosine Y_D is unaffected by PABG binding, while it is shifted slightly upfield when DAP binds. However, when both ligands are present the shift of this resonance is doubled.

By using n.m.r. spectroscopy in conjunction with binding and kinetic experiments, and in due course with crystallographic information as well, we can begin to define the conformational changes in structural terms and to assess their contribution to the overall binding energy.

Conclusions

The concept of antimetabolite action has been of very considerable value in guiding the design of new drugs over the last thirty years. With our increasing knowledge of cellular metabolism, one can design a new drug to interfere with a specific metabolic process by 'mimicking' the structure of a particular natural compound. However, there are limits to the extent to which one can use such a 'mimicking' process as the sole guideline for synthesis, as illustrated by trimethoprim and sulphadiazine (see above). Furthermore, the work on dihydrofolate reductase shows that even when, as in the case of methotrexate, the inhibitor is a close structural analogue of the substrate, they may not bind to the enzyme in precisely the same way. Thus we have been able to show that methotrexate binds in the protonated form, whereas folate binds as the neutral molecule, and in addition that the conformational changes which accompany binding are different in the two cases. Both these effects contribute to the difference in binding energy between folate and methotrexate. It is clearly an oversimplification to regard methotrexate simply as an analogue of folate which binds to the same site in the same way, and there is no reason to suppose that dihydrofolate reductase is an unusually complex drug receptor—rather the contrary. A proper understanding of the mode of action of antimetabolites thus requires that we study their interaction with their 'target' in considerable detail. Any rational attempt to design new and better antimetabolites must take into account the kind of complexity in the binding process revealed with dihydrofolate reductase, and this can only be done by studying drug action at the molecular level.

THE DEVELOPMENT OF THE CEPHALOSPORIN FAMILY OF ANTIBIOTICS

The cephalosporins are a family of β-lactam antibiotics, developed during the last 25 years, which now have wide use in medicine and annual world sales of more than £600 million. They are non-toxic bactericidal substances which resemble the penicillins in some of their properties, but differ from them strikingly in others. They may be regarded as complementary to the new semi-synthetic penicillins for the treatment of many serious bacterial infections and produce no reaction in most patients with penicillin allergy.

The first member of the family, cephalosporin C, was discovered in 1953. Like the first penicillin, this substance came from unexpected findings in the course of academically motivated research.

Historical

The story begins in 1945, when the chemotherapeutic properties of penicillin were first becoming known in Europe. Giuseppe Brotzu, a Sardinian professor of bacteriology, a former Rector of the University of Cagliari, a local politician and a man of affairs, decided to look in sea water near a sewage outfall for antibiotic-producing organisms. He isolated a strain of *Cephalosporium acremonium* which produced material with activity against a number of gram positive and gram negative bacteria; and he boldly injected crude extracts of this material into patients with a variety of infections, with results that he thought encouraging, particularly in the case of typhoid fever. Having tried, without success, to interest the Italian pharmaceutical industry and being unable to carry the work further himself, Brotzu published his findings in a unique issue of *Lavori dell' Istituto D'Igiene di Cagliari*, a journal that had never appeared before and has not done so since; and he kindly sent his fungus to Sir Howard Florey at the Sir William Dunn School of Pathology, Oxford, in 1948.

Penicillin N and Cephalosporin C

Work in Oxford soon showed that the Sardinian *Cephalosporium* produced at least two types of antibiotic. One, named cephalosporin P because it was active only against gram-positive bacteria, was shown to be a member of the steroid group. The second, which was first named cephalosporin N after it had been found to have activity against gram-negative bacteria, was labile and extremely hydrophilic. It first aroused our interest not because it was expected to have chemotherapeutic properties but because it appeared likely to be a peptide with a highly reactive chemical grouping, as was the antibiotic bacitracin on which work was then in progress. On further study by Newton and Abraham it turned out that cephalosporin N was a new penicillin, with a four-membered β-lactam ring fused with a thiazolidine ring, but with a δ-(D-α-aminoadipyl) side-chain in place of the phenylacetyl side-chain of the benzylpenicillin (I) in clinical use. Cephalosporin N was then renamed penicillin N.

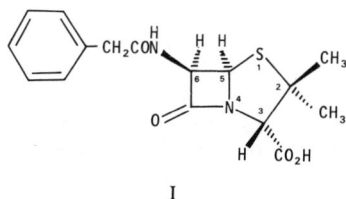

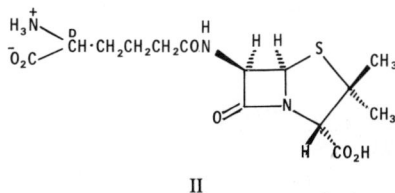

I   II

Penicillin N (II) was undoubtedly the substance whose activity had been observed by Brotzu. It was of interest as a potential chemotherapeutic agent because the α-aminoadipyl side-chain endowed it with a range of antibacterial activity quite different from that of benzylpenicillin. Morevoer, acylation of the amino group substantially increased its relatively low activity against the staphylococcus and decreased its activity against *Salmonella typhi*. Under the American name synnematin it was used in a small-scale clinical trial in typhoid fever and was reported to be more effective than chloramphenicol. But it never became available in quantity, probably because its isolation was too difficult and its use likely to be too limited for commercial production to be economically attractive.

However, it was the difficulty encountered in the purification of penicillin N that led to the discovery of cephalosporin C by Newton and Abraham in 1953. To obtain a pure compound for chemical investigation impure penicillin N was transformed   at   pH   3   into   its   isomeric   penillic   acid   (III,   R   =

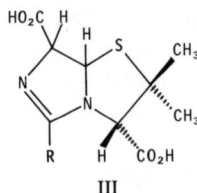

III

$^-O_2CCH(\overset{+}{N}H_3)(CH_2)_3-$), which had different physico-chemical properties. Chromatography on an anion exchange resin revealed that the penillic acid was mixed with a small amount of a compound showing an absorption band in the ultraviolet region with maximum absorption at 260 nm. This compound, named cephalosporin C, was readily isolated as a crystalline sodium salt. Preliminary studies indicated that it had a structural resemblance to penicillin N, but differed from the latter in part of its ring system, and that this difference endowed it with resistance to inactivation by a penicillinase (β-lactamase) which catalysed the hydrolysis of the penicillin β-lactam ring to give a (5R, 6R) penicilloate (IV).

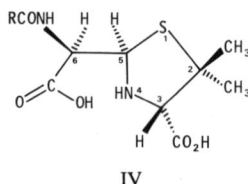

IV

These properties at once made cephalosporin C a compound of more than academic interest, because in 1953 a medical problem relating to the use of penicillin was becoming acute. Strains of staphylococci that were resistant to benzylpenicillin because they produced a penicillinase were beginning to predominate, particularly in hospitals where they grew selectively during use of the drug. There was thus an obvious need for an antibiotic with the desirable properties of penicillin, but, in addition, with resistance to destruction by staphylococcal penicillinase. Cephalosporin C, whose toxicity to mice was even lower than that of benzylpenicillin, was the first substance to show sign of meeting this need.

Structure and Properties of Cephalosporin C and its Nucleus

Chemical studies on cephalosporin C indicated that it contained sulphur and a fused β-lactam ring. In contrast to penicillin it gave no penicillamine (D-β-thiolvaline) on hydrolysis, but it yielded valine, as did penicillin, when hydrolysis was preceded by desulphurization with Raney nickel. Further work was carried out with material prepared at an Antibiotics Research Station of the Medical Research Council, where a higher-yielding mutant of *C. acremonium*

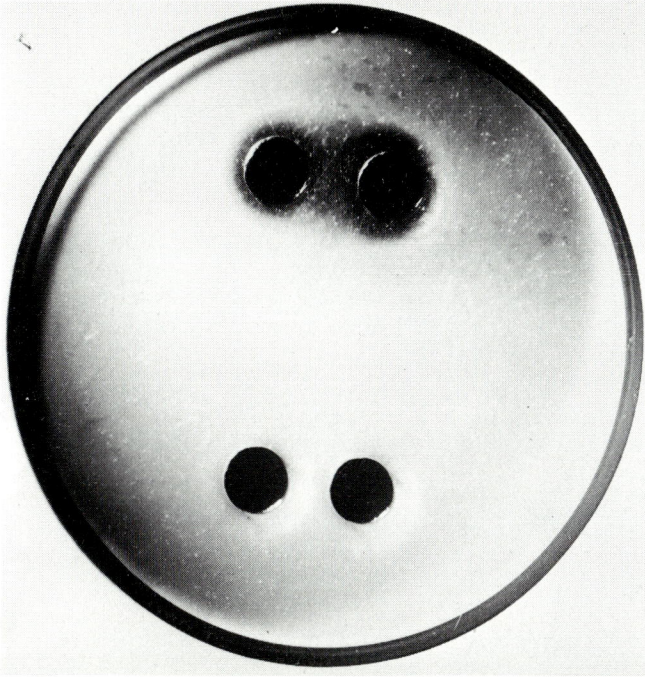

FIGURE 19.   Synergism shown by methicillin (top left) and cephalosporin (top right) on a plate seeded with a β-lactamase-producing strain of *Pseudomonas aeruginosa*. Cephalosporin C is hydrolysed by the β-lactamase and produces no inhibition of growth when used alone (bottom, left and right) but methicillin protects it from hydrolysis.

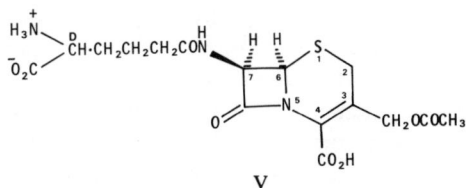

V

had been isolated, and finally with material from Glaxo. This led, in 1959, to the proposal by Abraham and Newton of structure (V), in which a β-lactam ring was fused with a dihydrothiazine ring instead of a thiazolidine ring. It could not have been predicted that a compound with this structure would have the ultraviolet absorption spectrum of cephalosporin C, but the structure was confirmed by an X-ray crystallographic analysis by Hodgkin and Maslen and the hydrogen atoms of the β-lactam ring (at C-6 and C-7 in V) were shown to be *cis*, as they are in penicillin. Drawings of models of the penicillin (A) and cephalosporin (B) ring systems are shown in figure 18.

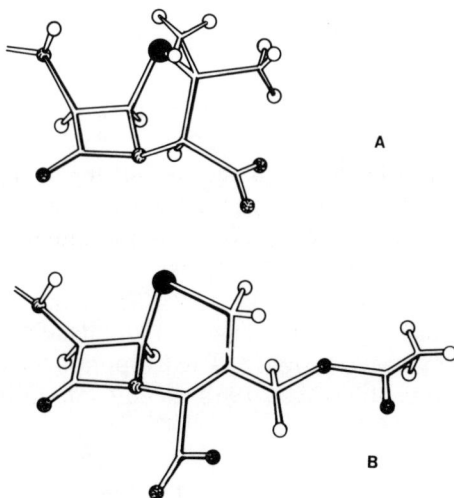

FIGURE 18. Drawings of wire models of the ring systems of (A) penicillins and (B) cephalosporins.

Had it not been for other developments at the end of the nineteen fifties, cephalosporin C would probably have been used in medicine for the treatment of infections with penicillinase-producing staphylococci. But the production in the Beecham Laboratories of the nucleus of the penicillin molecule, 6-aminopenicillanic acid, and then, by semi-synthesis, of a new and penicillinase-resistant penicillin, methicillin, made the use of cephalosporin C itself unlikely, because methicillin showed a higher intrinsic activity against the staphylococcus. However, argument by analogy from penicillin N and benzyl-penicillin suggested that exchange of the α-aminoadipyl side-chain of cephalosporin C for certain other side-chains, such as phenylacetyl, would yield penicillinase-resistant cephalosporins with a much higher intrinsic activity than the parent compound. This proved to be so. Loder, Newton and Abraham showed that mild acid hydrolysis under controlled conditions gave small amounts of 7-aminocephalosporanic acid (7-ACA, VI) and acylation of the latter with phenylacetyl chloride yielded a compound whose activity against the staphylococcus was more than one hundred times that of the parent cephalosporin C. In addition, it was found that the allylic acetoxy group of cephalosporin C (V) could readily be displaced by nucleophiles such as pyridine and that this type of displacement could also result in an increase in activity. On the other hand, hydrolysis of the O-acetyl group by an acetylesterase gave deacetylcephalosporins which were less active.

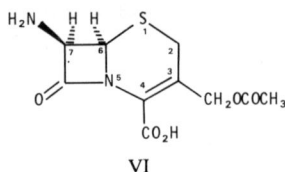

VI

These and other findings in Oxford gave us reason to hope that the cephalosporin (cephem) ring system would provide the nucleus of a series of new β-lactam antibiotics of medical value. The fulfilment of this hope was due in great measure to research workers in pharmaceutical companies in the USA, the UK, Switzerland and Japan, who exploited the situation with remarkable efficiency and made notable contributions, at the same time, to organic chemistry. These companies had negotiated agreements with the National Research Development Corporation, to which patents relating to cephalosporin C and its derivatives had been assigned. The Corporation had been set up under the Development of Inventions Act, 1948, to develop inventions in the public interest.

Semi-synthetic Cephalosporins

The production of 7-ACA in quantity was a major problem. By mild acid hydrolysis of cephalosporin C it was obtained in only very poor yield. Despite extensive searches, no enzyme was found which would remove the δ-(D-α-aminoadipyl) side-chain. The problem was first solved in the Lilly Research Laboratories. Treatment of cephalosporin C with nitrosyl chloride in formic acid converted the amino group of the side-chain into a diazonium ion and this led to the formation of an iminolactone which was readily hydrolysed to give 7-ACA. Another method, involving the conversion of the side-chain amide group to an imino-chloride, was subsequently introduced by CIBA–Geigy.

A further process of major importance was developed in the Lilly Research Laboratories. This was based on the finding that a penicillin (S)-sulphoxide (VII) could be brought into thermal equilibrium with an open-chain sulphenic acid (VIII) and that the latter could undergo ring closure to a deacetoxycephalosporin (IX). The way was thus opened to the production of deacetoxycephalosporins, and later of cephalosporins themselves, from penicillins; and compounds obtained by trapping the sulphenic acid (VIII) have since been widely used in the chemical synthesis of new β-lactam antibiotics.

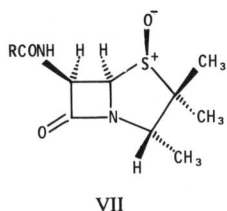

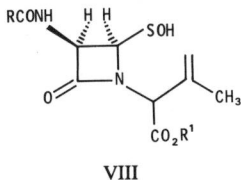

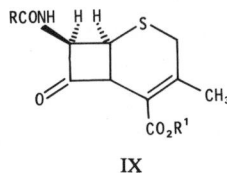

VII   VIII   IX

These developments led to the production in the 1960s of great numbers of cephalosporins with different side-chains and the general structure (X), in which R^1 and R^2 are variable. The first cephalosporins to be used in medicine (cephalothin, cephaloridine and cefazolin) could cope with infections with most gram-positive bacteria, including the penicillinase-producing staphylococcus, and certain gram-negative bacteria. Such compounds were not effective when given by mouth, but a deacetoxycephalosporin with a D-phenylglycyl side-chain (cephalexin) was subsequently found to be well absorbed from the gut.

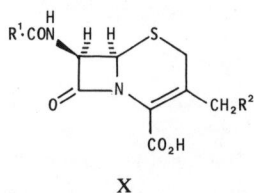

X

As the cephalosporins (and new penicillins) came into clinical use, changing patterns of infection focused more attention on the problems presented by gram-negative bacteria. Some of these organisms showed resistance to all the β-lactam antibiotics then available, and their resistance was due, at least in part, to the production of cell-bound β-lactamases that differed from the staphylococcal enzyme. Progress in the search for new cephalosporins that are resistant to hydrolysis by many of these β-lactamases has now been made by several pharmaceutical companies. In two cases this has come from the attachment of new side-chains to the cephalosporin ring system. In a third it has come from the discovery, in the Merck and Lilly Laboratories, that strains of the prokaryotic *Streptomyces* produce 7α-methoxycephalosporins (XI), with a

δ-(D-α-aminoadipyl) side-chain, whose β-lactam ring has increased resistance to enzymic hydrolysis. 7-Methoxycephalosporins with different side-chains have been obtained from these natural products and also from cephalosporins themselves by the use of chemical methods for the introduction of a 7 α-methoxyl group into the ring system.

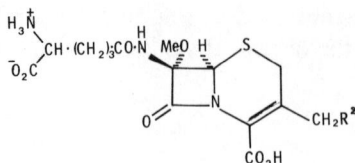

XI

Chemical Synthesis and Biosynthesis

A total chemical synthesis of cephalosporin C was first accomplished by Woodward, Heusler and others, at the Woodward Research Institute in Basle. The synthesis started from L-cysteine, which was used to build a key intermediate (XII) containing a β-lactam ring with the correct stereochemistry. This remarkable achievement has been followed by other syntheses in which the β-lactam ring has been formed by cyclic addition of an azidoketene to an imine.

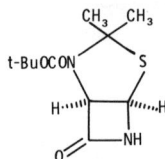

XII

Total chemical synthesis has never seemed likely to compete with semi-synthesis for the large-scale production of cephalosporins. Nevertheless, it has yielded substances with a modified ring system which have not so far been encountered as products of micro-organisms and these nuclear analogues have extended our knowledge of the changes that can be made in the ring system itself without substantial loss of activity. For example, replacement of the sulphur atom by oxygen or CH_2 and replacement of the methylene at $C-2$ by sulphur gave compounds with activities similar to that of the original cephalosporin. Such compounds provided the first demonstration that a sulphur-containing ring was not an essential component of β-lactam antibiotics. This finding suggested that further β-lactam antibiotics, with neither the penicillin nor the cephalosporin ring system, might be found in nature, since one hypothesis about the evolution of genes specifying antibiotic biosynthesis assumes that the antibacterial activity of these substances has been of advantage to the organisms producing them.

Within the last two years new compounds have in fact been found in the course of large-scale screening programmes in which bacteria used for testing activity are supersensitive to β-lactam antibiotics, or in which culture fluids have been tested for an ability to inhibit the action of β-lactamases rather than the growth of bacteria. The only common feature in the structures of these compounds and those of the cephalosporins and penicillins is a β-lactam ring.

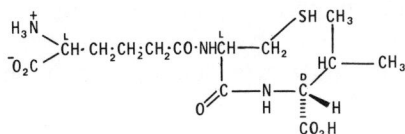

XIII

The biosynthesis of the cephalosporins is closely linked to that of the penicillins. A common intermediate appears to be the tripeptide δ-(L-α-aminoadipyl)-L-cysteinyl-D-valine (XIII), which yields isopenicillin N (with an L-α-aminoadipyl side-chain) in *P. chrysogenum* and penicillin N in *C. acremonium*. Isopenicillin N is converted by an acyl transferase to benzylpenicillin and penicillin N may be the precursor of deacetoxycephalosporin C. However, the biosynthesis of some of the new β-lactam compounds appears to start from different structural units and follow different pathways.

Bacterial Resistance and β-Lactamases

The cephalosporins, like the penicillins, inhibit the synthesis of bacterial cell walls by inactivation of enzymes involved in the cross-linking of peptidoglycan strands. Their activity depends on their ability to react with one or more of a number of binding proteins in the cell membrane, and thus on their ability to penetrate the cell wall. It also depends on their resistance to hydrolysis by a β-lactamase if such an enzyme is produced in significant amounts by the cell. Among the properties of the antibiotics which enable them to act as specific inhibitors or substrates of these different enzymes appear to be the stereochemistry of the ring systems and the reactivity of the fused β-lactam ring.

A β-lactamase which hydrolysed cephalosporin C and was described as a cephalosporinase, was first observed by Abraham and Newton in the culture of a gram-positive organism, *Bacillus cereus*, where it was accompanied by a second β-lactamase, mainly a penicillinase, of which cephalosporin C was a competitive inhibitor. Since then, at least twenty-five different β-lactamases, with different activities against the various members of the cephalosporin and penicillin families, have been found to be produced by gram-negative bacteria. These enzymes are cell-bound. Some of them are inducible and some are mediated by plasmids (extrachromosomal genetic elements) which can be transferred from one organism to another.

This multiplicity of β-lactamases has added greatly to the problems of chemotherapy by β-lactam antibiotics. Although progress has been made in the search for new cephalosporins with resistance to a wide range of enzymes, other approaches to the problem are possible. One depends on the ability of certain β-lactam compounds, which are not themselves hydrolysed at an appreciable rate, to combine with a β-lactamase and protect other compounds from hydrolysis. This can give rise to synergism, as shown in figure 19, plate I. A second depends on the finding that some β-lactam compounds deactivate β-lactamases (although the deactivation is often reversible) while they are undergoing hydrolysis. An illustration of this deactivation is provided by figure 20.

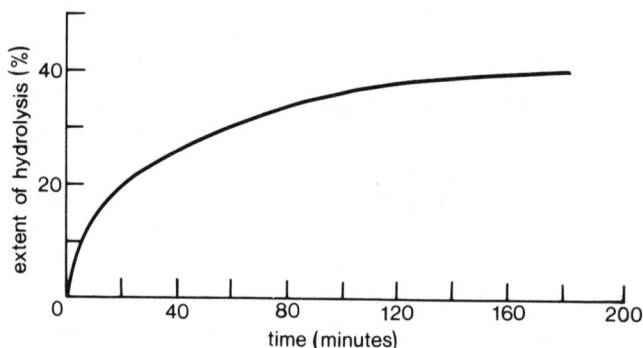

FIGURE 20. Deactivation of a β-lactamase (0.1nM) from *Pseudomonas aeruginosa* during hydrolysis of a cephalosporin (Glaxo 87/312, 0.2mM). (From Berks, D. Phil. Thesis, Oxford 1976.)

The present lack of understanding in molecular terms of the way in which β-lactam antibiotics react specifically with the active centres of β-lactamases and of certain proteins in the bacterial membrane is a major impediment to the rational design of new and useful substances. But a substantial increase in understanding can be expected in the next decade. Amino-acid sequences of four β-lactamases are now known, largely due to the work of Richard Ambler and his colleagues, and X-ray crystallographic analyses may reveal the three-dimensional structures of several β-lactamases in the forseeable future. Moreover, some of the enzymes in bacterial cell membranes that are sensitive to penicillins and cephalosporins are now becoming available for detailed investigation. Perhaps it is not too much to hope that the next 25 years will see the production of clinically effective compounds whose design was based on some knowledge of macromolecular structure.

HIGH BLOOD PRESSURE
THE EVOLUTION OF DRUG TREATMENT:
BRITISH CONTRIBUTION

W.D.M. PATON, F.R.S.[a], ELEANOR ZAIMIS[b], J.W. BLACK, F.R.S.[c]
and A.F. GREEN[d]

[a]University of Oxford
[b]Royal Free Hospital School of Medicine, London
[c]University College London
[d]The Wellcome Research Laboratories, Beckenham, Kent

The account which follows is based on the exhibit at the Silver Jubilee Exhibition. It must be emphasized that this was deliberately selective, picking out three particular drugs to illustrate an evolution, and adding some statistical data to provide context. The whole story, still developing, is a vast one, and much of it can be traced clinically in the books and papers by Sir John McMichael, Sir George Pickering and Sir Horace Smirk and their colleagues, while for an introduction to the physiology of the autonomic nervous system, one can do no better than to read Sir Henry Dale's *Adventures in Physiology* (1953).

THE CONTROL OF THE NORMAL AND RAISED BLOOD PRESSURE

The whole body comes under control of the brain through:
(1) the motor nerves carrying impulses from the neuraxis to voluntary muscle: voluntary (striated) muscle is not spontaneously active, and its movement is precisely controlled by the rate of discharge of impulses down the various motor nerves;
(2) the two divisions of the autonomic nervous system (sometimes called the 'vegetative' nervous system) which carry nerve impulses to organs not under voluntary control: 'smooth' muscle, heart and glands (figure 1); these are often spontaneously active and the activity can be modulated in either direction by sympathetic and parasympathetic nerves; for instance the sympathetic nerves accelerate the heart, the vagus nerve (parasympathetic) slows it.

We are here particularly concerned with one part of the sympathetic division, the *adrenergic neurone*. Classical work on the autonomic nervous system, and on the theory of chemical transmission of nerve effects, has shown that there are two chemical stages in the sympathetic outflow: (1) at the ganglionic relay

223

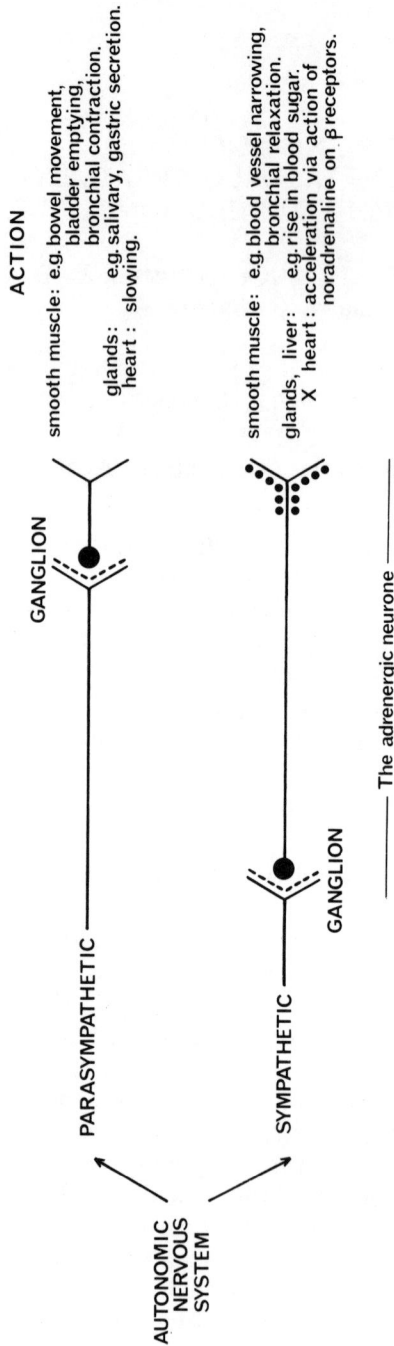

FIGURE 1.   Diagram to show the divisions of the autonomic nervous system, the adrenergic neurone, and the evolution of the progressively selective attack on adrenergic function. – – – site of ganglion block, leading to reduction of both parasympathetic and sympathetic function; ●●● site of adrenergic neurone block, leading to general reduction of sympathetic activity; X site of β-adrenergic block, leading to reduction of sympathetic effect on the heart.

acetylcholine is released to carry the excitation to the next neurone; (2) at the nerve-endings in the effector organ, noradrenaline is released to produce a final chemical stimulus. The neurone is therefore called 'adrenergic'.

Reduction of the activity of the adrenergic neurone, or of the effect of the final transmitter, noradrenaline, has been the main basis for the treatment of hypertension. The demonstration illustrated the use of ganglion-block, prevention of noradrenaline release, or selective block of noradrenaline action for this purpose. But it must be emphasized that there have been many other contributions: e.g. reserpine (which reduces the amount of noradrenaline present); veratrine (which produces a reflex reduction of sympathetic activity); methyldopa (which acts principally by generating a 'false transmitter' at central adrenergic synapses). One important exception to this autonomic attack is the use of thiazide derivatives. These cause the body to lose salt and water (although this cannot account for all their action) and augment the effect of other hypotensive drugs in a clinically important way.

When the blood pressure is consistently maintained at too high a level—hypertension—the smooth muscle in the arterial wall thickens and the heart hypertrophies. The arterial change can then lead to further damage, particularly in three target areas—eyes, brain and kidney; and a characteristic type of heart failure may develop.

The introduction of treatment with hexamethonium quickly showed that reducing the blood pressure:

(a) relieved the severe headache of malignant hypertension;

(b) allowed retinal damage to regress, with improvement of vision (figure 2, plate I);

(c) allowed the enlarged heart and oedema of the lung in hypertensive heart failure to return towards normal (figure 3, plate II);

(d) allowed the patient to return to work.

In due course, it became clear that the expectation of life of the malignant hypertensive could be considerably improved. It had, in fact, long been thought that high blood pressure was damaging to blood vessels. But there was also a fear that the raised blood pressure might be compensating for the narrowing of the blood vessels through thickening of their walls, so that lowering blood pressure might deprive the heart of essential blood supply.

The important point at this stage was the recognition that, *provided the arteries were not too damaged, a fall in blood pressure could be obtained by suitable reduction of sympathetic activity, without loss of tissue blood-flow,* because the blood vessels dilated with the removal of sympathetic stimulation. Acceptance of this conception was helped by the development by Enderby in 1951 of the use of ganglion-blocking agents to lower the blood pressure at surgical operation, in order to reduce bleeding. It was shown to be possible, for instance, to enable a plastic surgeon to operate on a nearly bloodless field, while blood flow in vital organs was maintained. Postural hypotension, liable to be a nuisance in the treatment of high blood pressure in ambulatory patients, was here usefully exploited.

HEXAMETHONIUM

Because the transmission of nervous activity across the ganglionic relay is chemical, mediated by acetylcholine, it becomes susceptible to chemical attack by the ganglion-blocking drug hexamethonium, as shown by Paton and Zaimis in 1948. Hexamethonium has a structural resemblance to acetylcholine (figure 4), and is able to react with receptors in the ganglion without activating the cells.

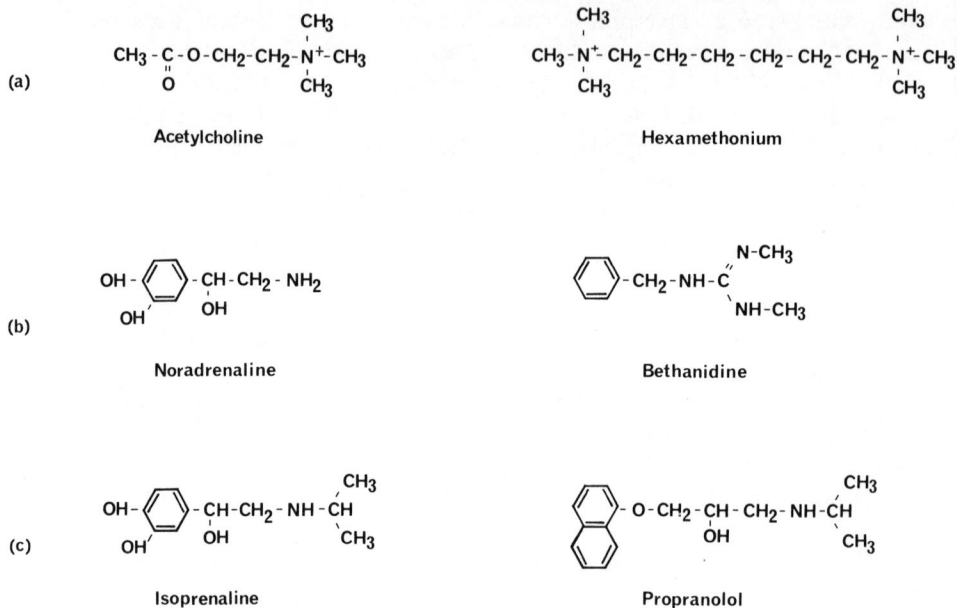

FIGURE 4. Chemical formulae showing the relationship in chemical structure between (a) the neurotransmitter acetylcholine and the ganglion-blocking agent hexamethonium; (b) the neurotransmitter noradrenaline and the adrenergic neurone blocking agent, bethanidine; (c) the β-stimulant isoprenaline and the β-blocker, propranolol. The similarity is more obvious in three-dimensional models.

It thus 'competes' with the natural transmitter, which cannot then produce its usual effect. The block of transmission is reversible, and acetylcholine release by the nerve ending is unimpaired. One can readily show, for instance, that action potentials in the nerve leading to the ganglion are unaffected by hexamethonium, while action potentials in the nerve leaving it are progressively and rapidly reduced or abolished as the block develops. Figure 5 gives an example of the fall in blood pressure produced by hexamethonium in a severely hypertensive patient. Figures 2 and 3 show the clinical improvement which results from such treatment.

BETHANIDINE AND THE INHIBITION OF NORADRENALINE RELEASE

A number of compounds were discovered from 1954 onwards that selectively inhibit the release of the adrenergic transmitter noradrenaline, of which

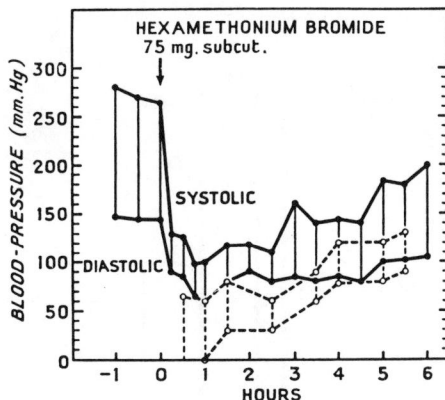

FIGURE 5. Response of blood-pressure to sub-
cutaneous hexamethonium after 16 months' treatment: continuous line, patient lying down;
interrupted line, patient standing up.
    (From Harington, M. and Rosenheim, M.L.,
Lancet **1**, 7–13, 1954.)

bethanidine proved the most satisfactory (Hey and Willie, xylocholine, 1954;
Boura and Green, bretylium, 1959; bethanidine, 1963). This inhibition can be
demonstrated directly by using a method developed by Sir Lindor Brown and
Professor Gillespie in 1957 measuring the amount of noradrenaline released
from a cat spleen into the venous blood by a train of electrical impulses applied to
the sympathetic nerve. Physiologically, the noradrenaline released makes the
capsule of the spleen contract, emptying its reserve of blood into the circulation
when needed. Figure 6 shows the effect of bethanidine in reducing noradrenaline
release; with increase of dose, release can be virtually abolished.

Drugs of this kind, termed 'adrenergic blocking agents', suppress responses
of the heart and blood vessels that are mediated by adrenergic nerves, but not the
corresponding responses to injected noradrenaline or reflexes mediated by
parasympathetic nerves such as the vagus, because neither noradrenaline nor
acetylcholine receptors are interfered with. Figure 7 illustrates this on the heart
rate of the rat. Bethanidine greatly reduces the increase in heart rate to
sympathetic stimulation, but leaves untouched the increase with noradrenaline
and the slowing to vagal stimulation.

Our knowledge of the mechanism of the release of noradrenaline (NA) from
adrenergic nerves has increased by studying these and other drugs. Gaps in our
knowledge remain, but some of the events are represented in figure 8, showing
one of the characteristic swellings ('varicosities') of the terminal branches of an
adrenergic nerve close to an effector cell. Figure 9 (plate III) is an electron
microscope picture of such a varicosity between two smooth muscle cells. The
noradrenaline is contained in granules, and the contents of these are released
from the cell by a special process called exocytosis; the noradrenaline can then
diffuse across the narrow gap to the effector cell to interact with receptors there.

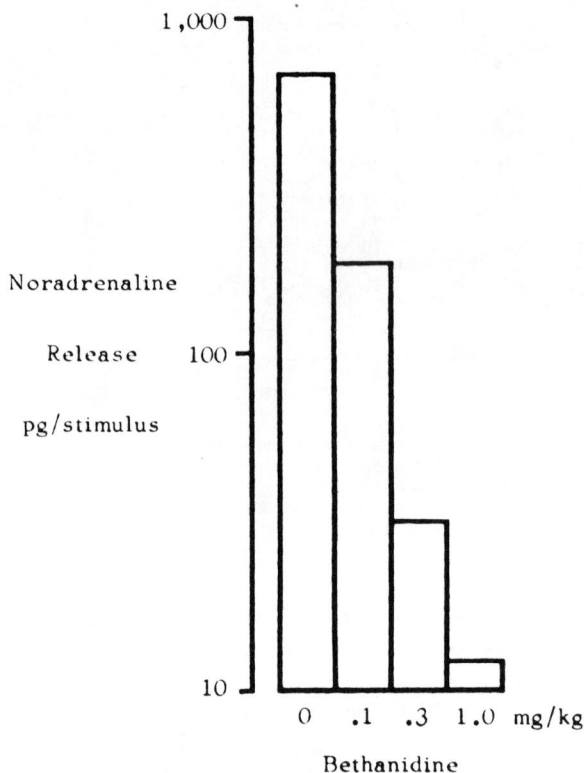

FIGURE 6.   The noradrenaline released per stimulus derived from the noradrenaline content of the venous blood from the spleen during stimulation of the splenic nerve in an anaesthetized cat. The noradrenaline released decreases with increasing dosage of bethanidine. (From original data referred to by Boura and Green, *Brit. J. Pharmac.* **20**, 36–55, 1963.)

A considerable part of the noradrenaline is then recovered by the nerve terminal, thus economizing in the amount of transmitter that needs to be synthesized. It seems that it is a similar process which leads to uptake of bethanidine by the nerve terminal, leading to a block of release. Figure 10 shows how bethanidine (radioactively labelled so that the amount present in tissue can be measured) is taken up much more readily by adrenergic nerves and by the ganglion from which they originate than by other nervous tissues.

Adrenergic neurone-blocking agents superseded ganglion-blocking agents for the control of hypertension in man because of their greater selectivity of action; the unwanted effects of parasympathetic blockade were avoided. Of such agents, bethanidine has the advantage of being completely and reliably absorbed when given orally. Its persistence of effect, illustrated in figure 11, allows continuous control of hypertension with oral doses of about 10 mg three times a day, while permitting rapid withdrawal of the drug action if required.

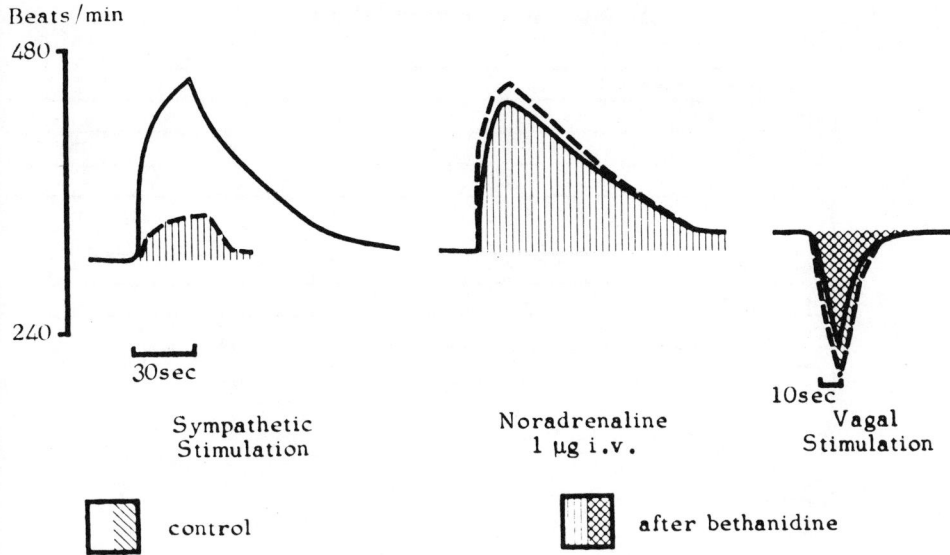

FIGURE 7. Illustration of the effect of bethanidine (0.4 mg kg^{-1} i.v.) on the rate response of the heart in a rat prepared as described by Gillespie, Maclaren and Pollock (*Brit. J. Pharmac.* **40**, 257–267, 1970) to stimulation of its sympathetic and its vagal nerve supply and to injection of noradrenaline.

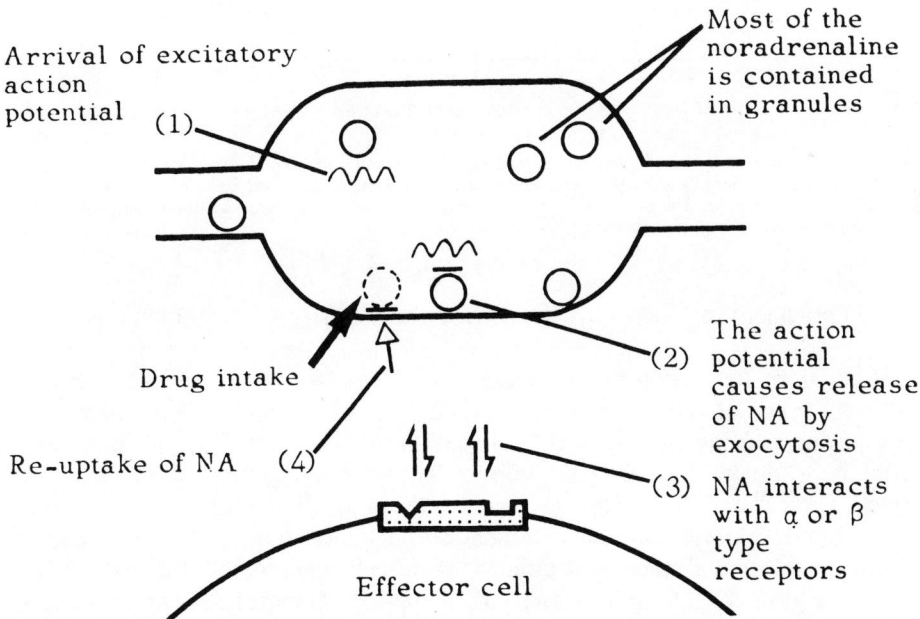

FIGURE 8. Diagram of part of an adrenergic nerve-ending (above) in relation to its effector cell (below), representing the release and re-uptake of the transmitter noradrenaline (NA), its reaction with α or β receptors on the effector cell, and the possibility of uptake of a drug such as bethanidine and interference with exocytosis.

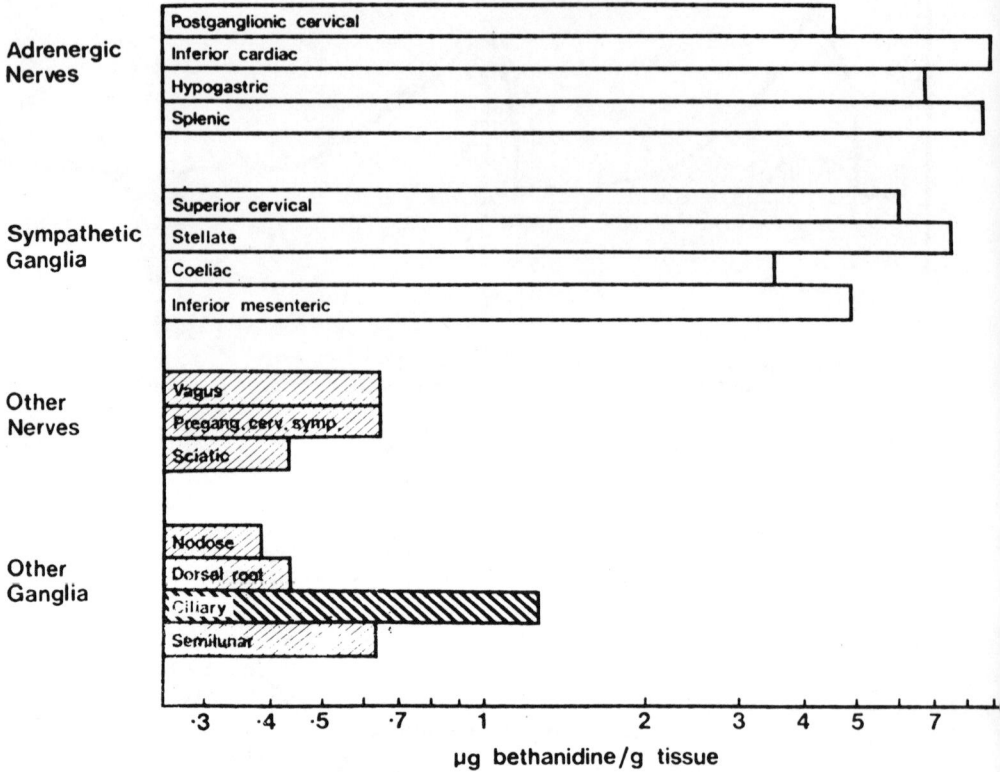

FIGURE 10.   The concentration of ^{14}C-labelled bethanidine in peripheral nervous tissue of cats at 16 hr after giving N-(α^{14}C) benzyl labelled drug (3 mg kg^{-1} s.c.). (From data of Boura, Duncombe, Robson and McCoubrey, *J. Phar Pharmac.* **14**, 722–726, 1962). The presence of adrenergic nerves running uninterrupted through the ciliary ganglia m account for their relatively high concentration of bethanidine.

PROPRANOLOL AND COMPETITIVE ANTAGONISM AT β RECEPTORS

It has been found that noradrenaline and drugs like it can interact with two different types of receptors, the so-called α and β receptors. The distinction arose from studying the patterns of action of the various stimulant ('agonist') and blocking ('antagonist') drugs. From selective β-antagonists discovered by Black, Duncan and Shanks in 1962 evolved the clinically important propranolol. Table 1 gives a simple summary of the position, concentrating on the β receptor. These receptors are macromolecular sites on cell membranes (for instance of heart or arterial muscle cells) which can 'recognize' noradrenaline molecules. A brief combination between noradrenaline and a β receptor switches on an adjacent enzyme leading to a change in cellular activity. Propranolol is also 'recognized' by β receptors and so can compete with noradrenaline for their occupation. When this molecular combination takes place, the specific enzyme is not switched on but at the same time a noradrenaline molecule is excluded. The likelihood that noradrenaline molecules in random motion will hit some β receptors is determined by the product of the concentration of noradrenaline and the concentration of unoccupied receptors. Therefore if, say, half the receptors are already occupied by propranolol, the same product, the same likelihood and thus the same effect on the cells as before can be had if the concentration of noradrenaline is simply doubled. Figure 12 illustrates this type of phenomenon in man. Here the drug isoprenaline (isoproterenol is the American name) was infused, and the heart rate increased as the dose decreased. Then the subject was treated with the antagonist propranolol, and the experiment repeated. It was found that the amount by which the dose of isoprenaline had to be increased to produce a given rise in heart rate was about ten-fold, regardless of the magnitude of the rise chosen. This 'parallel shift of a log-dose-response curve' is good evidence that propranolol competes with isoprenaline for the same receptors.

TABLE 1

Diagram of the Contrasting Relationships between
α and β Adrenergic Receptors

Receptors	α	β
Neurotransmitter	Noradrenaline	Noradrenaline
Selective agonist	–	Isoprenaline

Actions	Contraction of muscle in:
	  blood vessels
	  gut sphincters
	  eye
		Relaxation of muscle in:
		  blood vessels
		  gut wall
		  bronchi
		Stimulation of heart

Selective antagonist	–	Propranolol

The fact that propranolol treatment would lower the blood pressure was not predicted. In fact, one would expect it to raise blood pressure slightly, because noradrenaline can relax some blood vessels by an action on β receptors, tending to reduce its overall pressor effect; and in fact, with single intravenous doses of propranolol one indeed finds a constriction of blood vessels and a rise in blood pressure (figure 13). But deeper analysis shows a subtler picture. Figure 14 shows how, in a human subject, the rise in heart rate on exercise (which is an effect mediated by β receptors) is associated with a rise in blood pressure: so that the effect of a β-blocker during exercise at least, might be different.

In the event, it was noticed by B.N.C. Prichard during treatment of patients with propranolol for another purpose (angina), that as treatment continued, the blood pressure slowly fell. The response is different from that to ganglion-block by hexamethonium or to adrenergic neurone block by bethanidine, in that, as well as being more selective, it comes on over days or even weeks. The

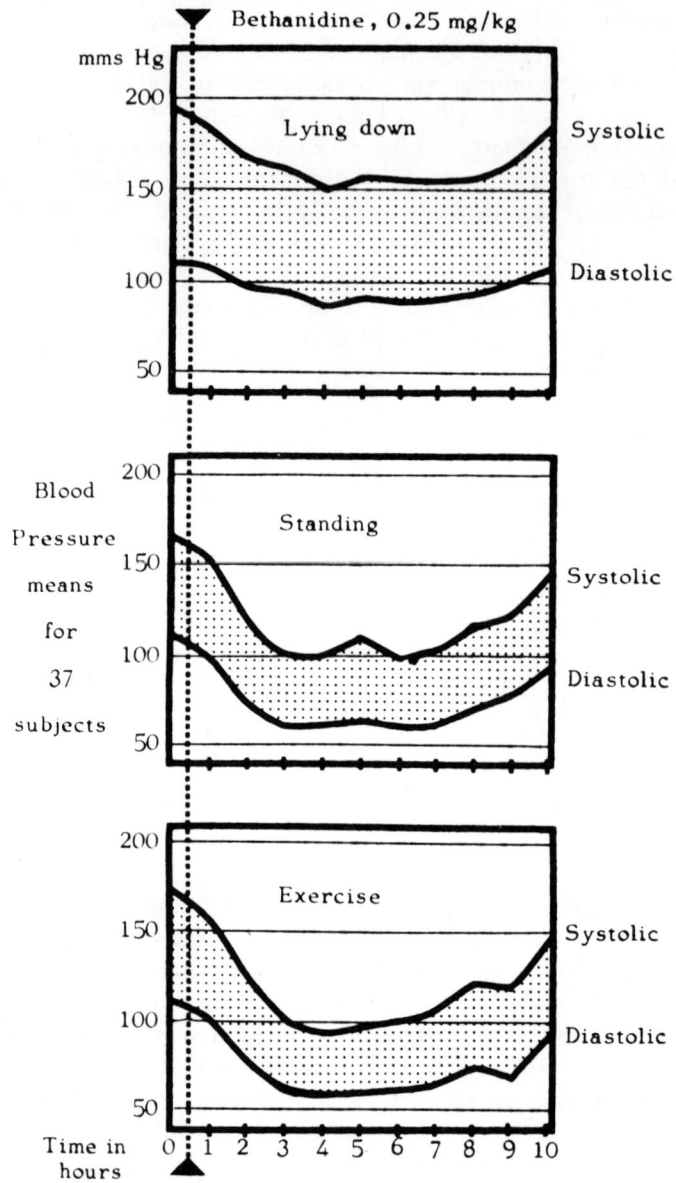

FIGURE 11. Lowering of blood pressure by bethanidine in hypertensive subjects, lying down, standing or during exercise. (From Wilson, R., Long, C. and Jagoe, W.S., *J. Irish med. Assoc.* **61**/331, 9–11, 1965.)

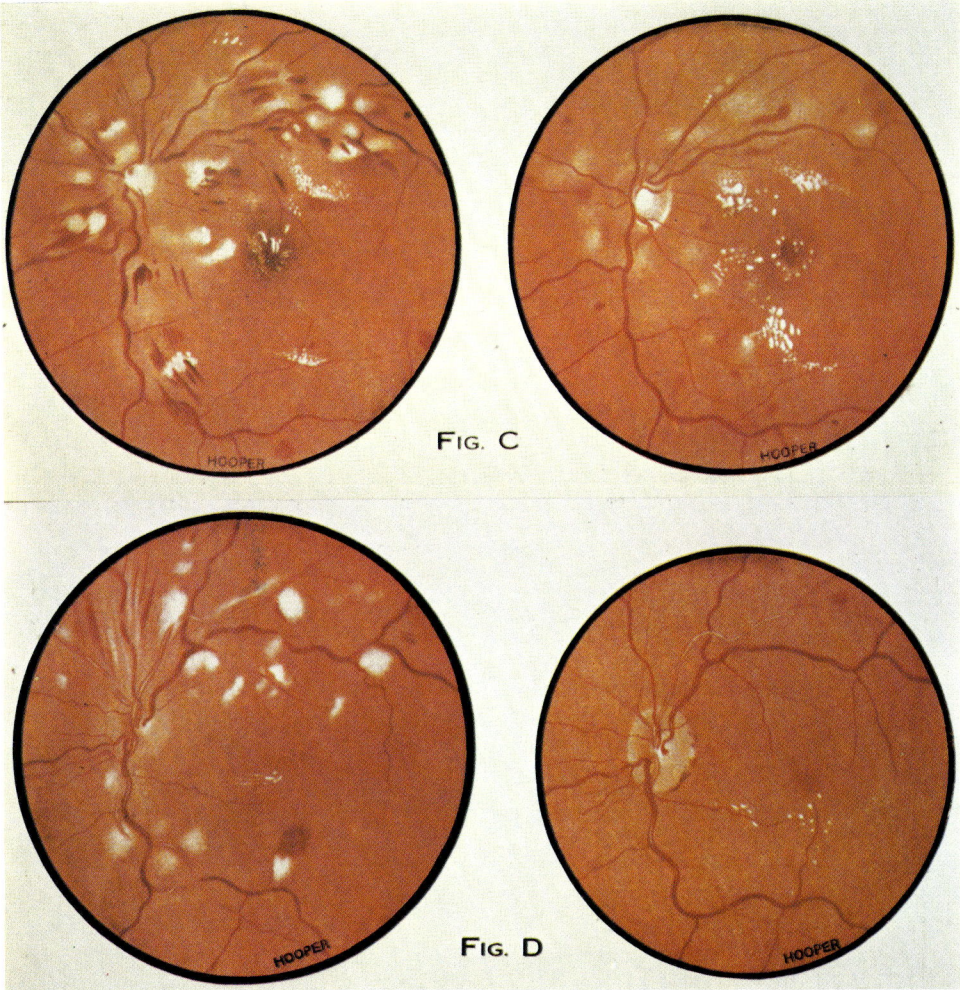

Figure 2.   Upper: Drawings of the fundus of a man aged 26 with chronic nephritis and uraemia before and after three months' treatment with hexamethonium.

Lower: Drawings of the fundus of a man aged 52 with malignant essential hypertension before and after seven months' treatment with hexamethonium.

(From Rosenheim, M.L., *Brit. med. J.* **2**, 1181–93, 1954.)

PLATE II (PATON *et al.*)

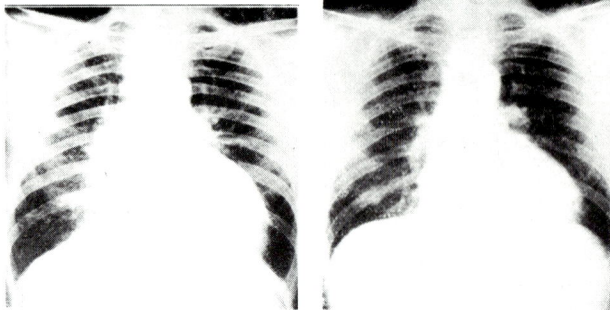

Before   After

Male aged 45. The x-ray film prior
to treatment was taken August 28, 1951,
and corresponded with mild congestive
heart failure and attacks of cardiac
asthma.   These cleared on methonium
treatment without other measures.   The
patient returned to work.   Second film
was taken May 4, 1953.

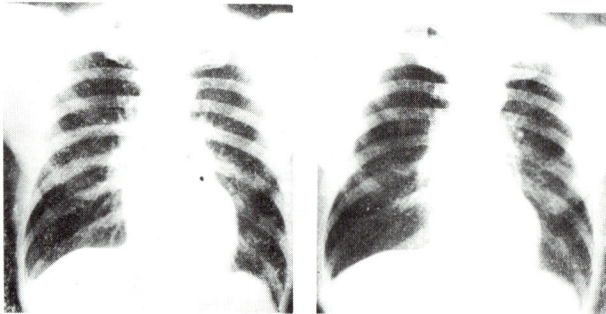

Before   After

Male aged 46. Before treatment there
was mild breathlessness but no congestive
failure or cardiac asthma. The first
x-ray film was taken on January 27, 1950.
Breathlessness cleared on methonium
treatment.   The second film was taken
on June 12, 1953.

FIGURE 3.   Chest X-rays of two hypertensive patients before and after
treatment by ganglion-block showing the reduction in size of the heart.
(From Smirk, *Brit. med. J.* **1**, 717, 1954.)

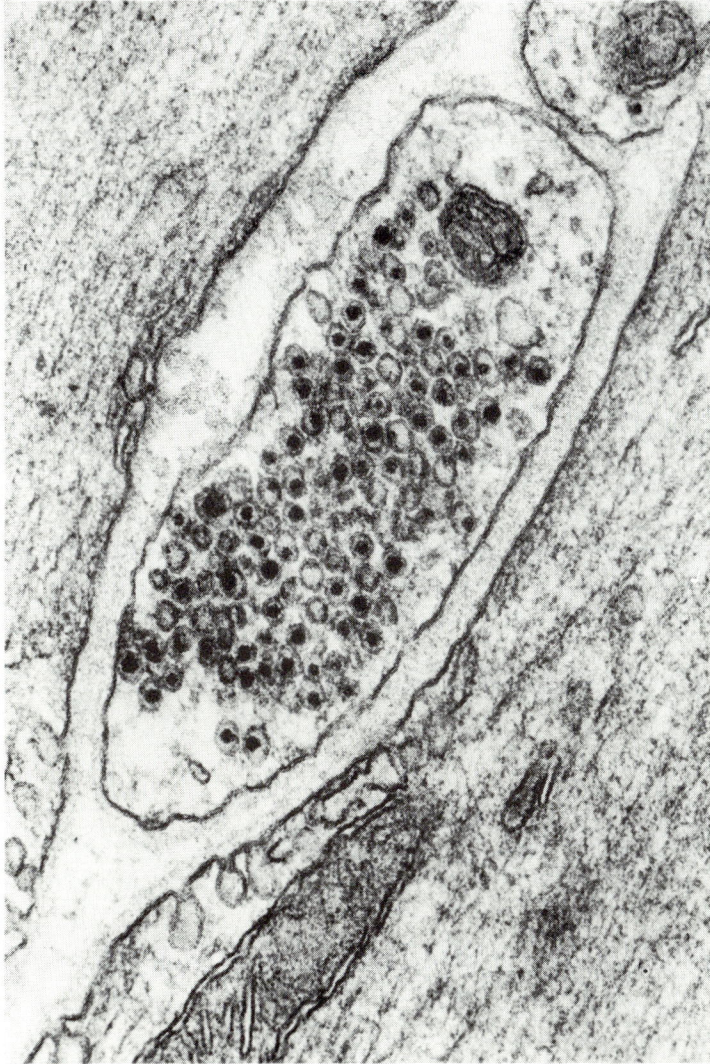

Figure 9. Electron micrograph of an adrenergic terminal (centre) with its granules containing noradrenaline, between two smooth muscle cells cut nearly longitudinally. (By courtesy of Professor G. Burnstock.)

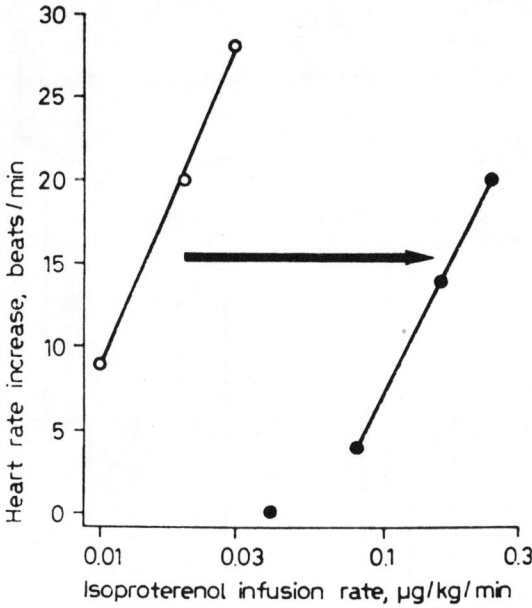

FIGURE 12. Parallel displacement of isoproterenol (isoprenaline) dose-response curves in a patient treated with
propranolol ⊖ Before, ● during propranolol 40 mg 4 times
daily. (From Zacest, R. and Koch-Weser. J., *Pharmacology* 7,
178–184, 1972.)

mechanism of the response is still debatable, but at least it is established that it
depends on β-blocking action.

One fascinating aspect is that under propranolol the blood pressure becomes
much less sensitive to posture than with other drugs active on sympathetic
mechanisms. Figure 15 illustrates this from a patient who transferred gradually
from bethanidine to propranolol. Under bethanidine, systolic and diastolic blood
pressures were much lower (130/80) when standing than when supine
(213/137), but when the transfer was over, it can be seen that the blood pressures
were about 145/90, regardless of position.

THE OUTCOME OF HYPOTENSIVE THERAPY

It soon became clear, with the early treatment of hypertension by
ganglion-block, that as well as lowering the blood pressure and relieving
symptoms, life could be prolonged. Figure 16 shows the survival of 140 patients
with the severe form of hypertension called 'malignant' (in which the eyes and
kidneys become seriously damaged) compared with the fate of 105 similar
patients from earlier years who had not been treated. The expectation of life
increased 6–8 times. Although a 'double-blind' clinical trial would have been

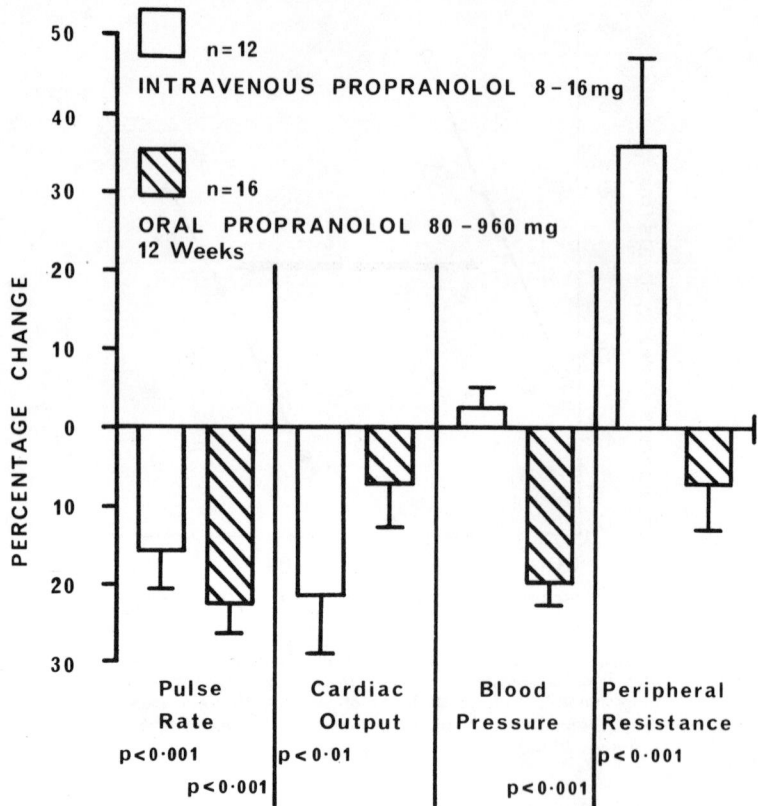

Figure 13.   The fall in blood pressure and decrease in peripheral resistance do not take place immediately the β-receptors are blocked so that this effect of continued treatment by propranolol was unexpected. (From Joekes, A.M., Thompson, F.D. and Girling, M., in: *Hypertension—its Nature and Treatment*, CIBA, England, 1975, pp. 181–186.)

more satisfactory, the early amelioration of heart failure and blindness by this treatment ruled out this method for malignant hypertension. But the magnitude of the benefit meant that this form of therapy was important, and the side-effects of ganglion-block provided a stimulus to discovering better methods.

Figure 17 shows the results of controlled clinical trials about ten years later using later methods of controlling hypertension which have less side effects and could thus be used for less severe cases for whom it was not established that treatment would be beneficial. It was found that there were fewer deaths and much fewer complications (e.g. stroke) in the treated group. Figure 18 develops this latter point. In a further study, some years later, the comparison was made between the success in controlling blood pressure and the incidence of particular complications (stroke, cardiac infarction, angina or heart failure), both in patients who had already had one stroke and in others who had not. This showed how good control of blood pressure greatly reduced the incidence of stroke and of heart failure; but it had no significant effect on the incidence of heart attack and angina.

Finally, one can ask if national statistics reflect these developments. The question is a difficult one, because the composition of the population changes over the years. This can be dealt with by using the age and sex specific death rates calculated in the Registrar-General's Reviews. A second difficulty is that the classification of diseases changes over the years. Fortunately, over the transition periods, data are calculated both by old and new methods so that equivalent classifications can be followed. Unfortunately age and sex specific data were not available for hypertensive disease before 1950; accordingly total death rates for hypertensive disease have been extracted over the whole period. Figure 19 shows the trends thus revealed. Since hypotensive drug treatment began, from 1949 onwards, there has been a downward trend in hypertensive deaths, somewhat greater in women. There is also an interesting confirmation of the Chelmsford study, namely an indication of a slow decline in deaths from strokes (cerebrovascular disease), especially in women, but no change in the steadily rising trend in deaths from ischaemic heart disease.

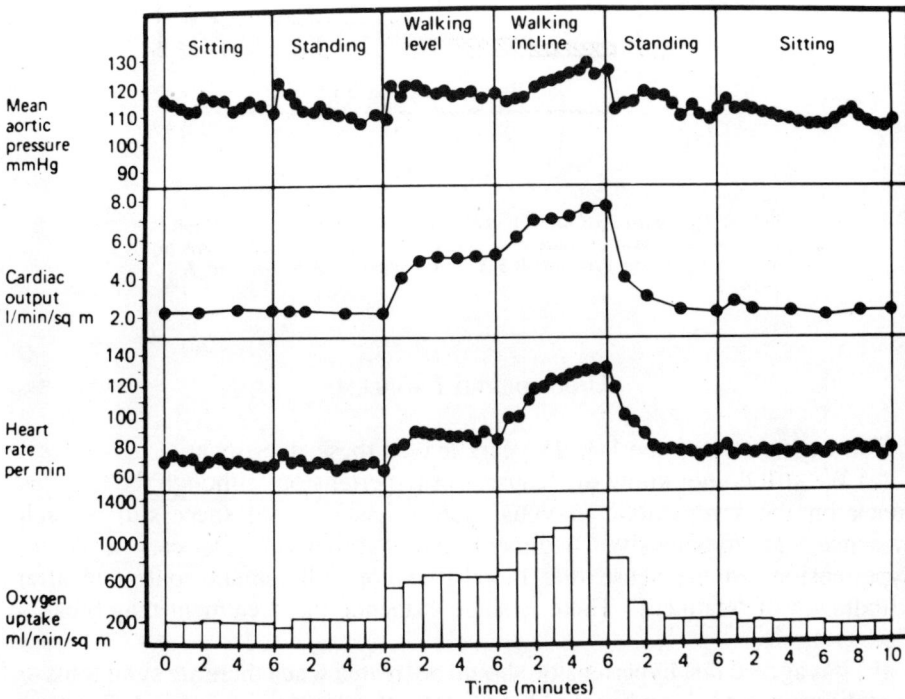

FIGURE 14.   Circulatory changes associated with walking in a healthy male aged 52 years. Enhanced sympathetic activity increases heart rate and cardiac output and raises blood pressure. This rise in blood pressure would be relatively greater in a patient with uncomplicated hypertension. (From Taylor, S.H. in: *Hypertension—its Nature and Treatment*, CIBA, England, 1975, pp. 29–53.)

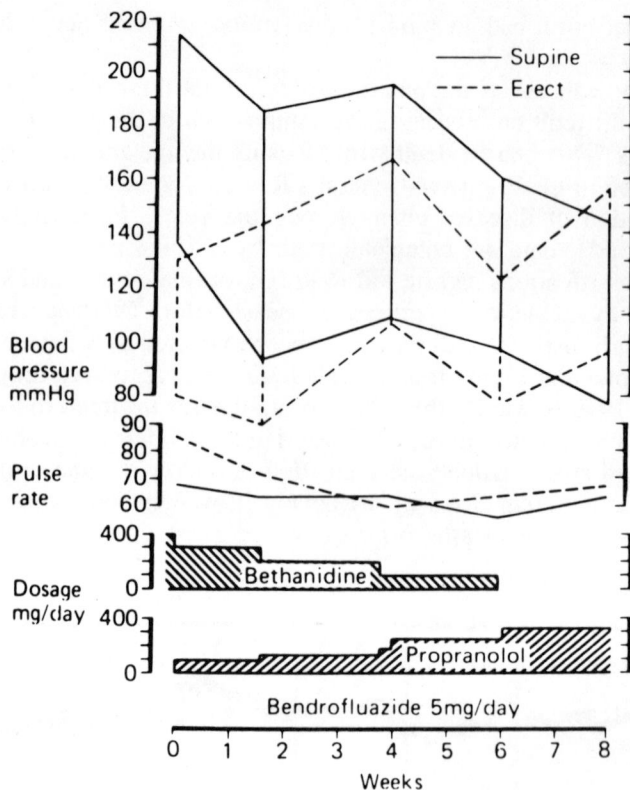

FIGURE 15.   Man aged 43 with essential hypertension. Record of supine and standing blood pressure over period of change from bethanidine to propranolol. (From Prichard, B.N.C. and Gillam, P.M.S. *Br. med. J.* 1, 7–16, 1969.)

UNRESOLVED PROBLEMS

The experience of the last 25 years leaves these questions:

(a) We still do not know the 'causes' of hypertension, although the fact that attack on the sympathetic nervous system has been so successful is itself evidence. Is hypotensive treatment only palliative? Or can it 'cure' hypertension, in the sense that blood-pressure will remain controlled after withdrawal of treatment? There is some evidence that treatment may break a vicious circle.

(b) It is agreed that hypertension should be treated when there are symptoms or complications and also when the blood pressure is about a certain level. But there is uncertainty as to precisely where that level lies. In trying to decide how extensive treatment should be, a balance must be struck between

(1) the expected benefit to the patient (who will be symptomless at the start of

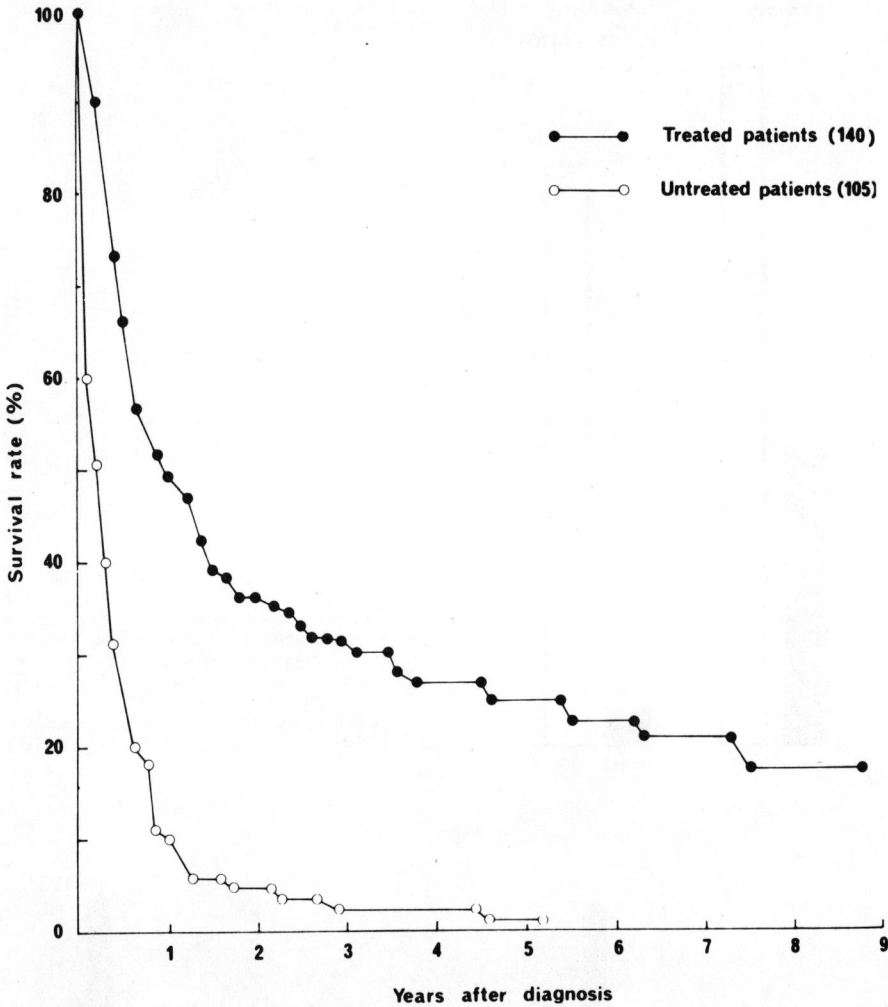

FIGURE 16.   Comparison of the survival of 140 patients with malignant hypertension treated with ganglion-blocking drugs with survival of 105 untreated patients. (From Harington, M., Kincaid-Smith, P. and McMichael, J., *Br. med. J.* 2, 969–980, 1959; and Harington, M., *Proc. roy. Soc. Med.* 55, 283–286, 1962.)

treatment) in warding off complications, together with the resultant saving in health service costs; and

(2) the cost of identifying (by screening) the individuals to be treated, and of providing treatment, together with the cost to the patient of any adverse reactions to the drugs over a life-time of treatment. This is a controversial question, involving problems of ethics, cost–benefit analysis, and organization.

(c) What is the mechanism of the gradual, unexpected, beneficial effect of propranolol? Its discovery opens up new areas of long-term neuro-circulatory

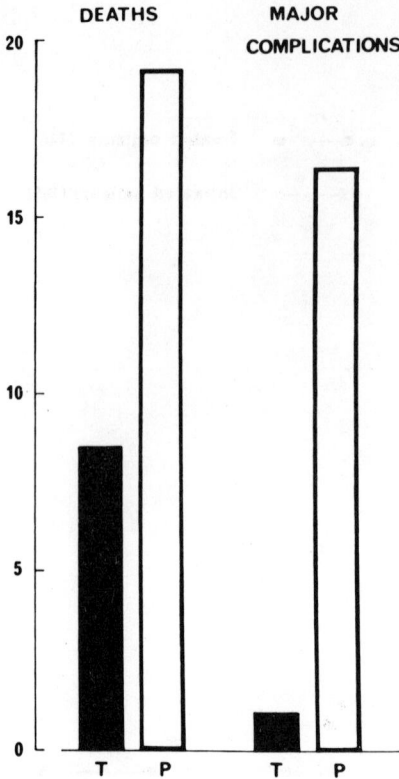

FIGURE 17. Controlled clinical trial of hypotensive therapy vs. placebo by US Veterans Administration, in hypertension without complications at entry into trial. 380 males, 2½ years follow-up, diastolic blood pressure 90–114 mmHg. T = treated; P = placebo. (Adapted from Dollery, C.T., *Br. med. Bull.* **29**, 158–162, 1973, using data from *J. Amer. med. Ass.* **213**, 1143–1152, 1970.)

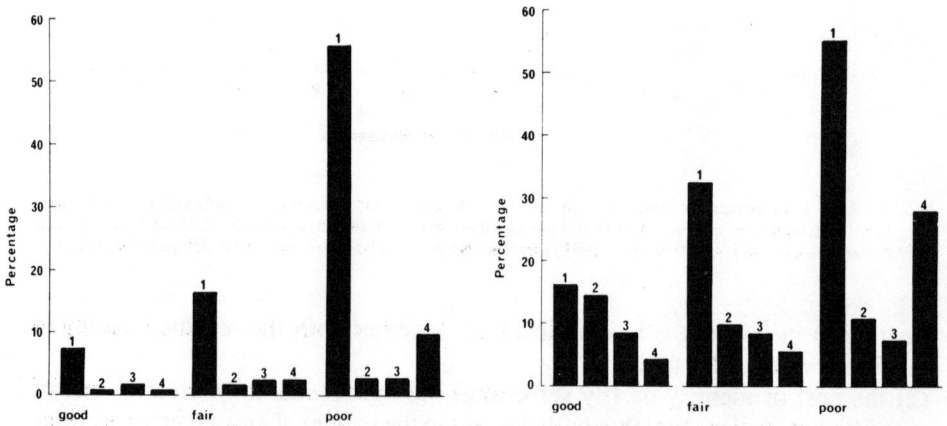

FIGURE 18. Comparison of clinical progress in relation to success in controlling blood pressure (Chelmsford trial). 1, stroke; 2, myocardial infarction; 3, angina; 4, cardiac failure. Right: 162 patients who had already had one stroke before treatment began. Left: 499 patients with no clinical signs of cerebral vascular disease before starting treatment. Note that with successful control of blood pressure cardiac failure and strokes were lower, but angina and myocardial infarction were unaffected. (Adapted from Hamilton, M. in: *Topics in Therapeutics*, ed. Shanks, R.G., Pitman Medical Co., 1977, pp 180–188.)

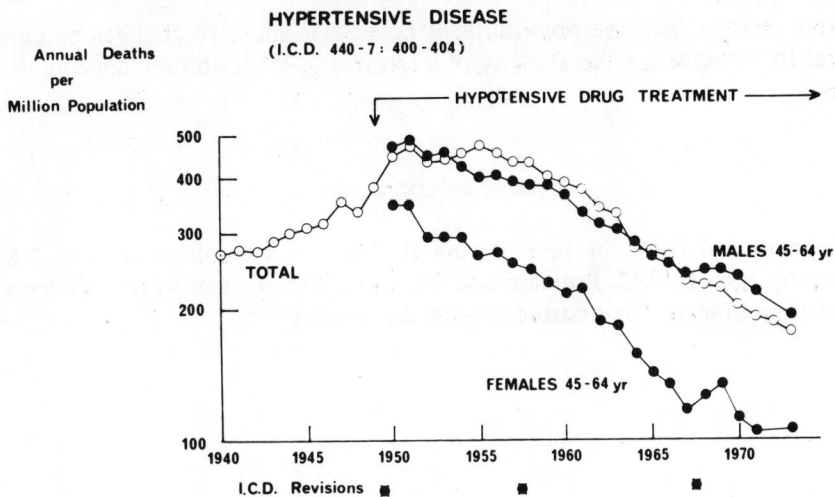

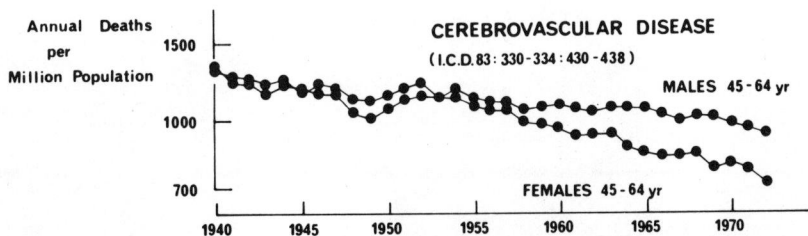

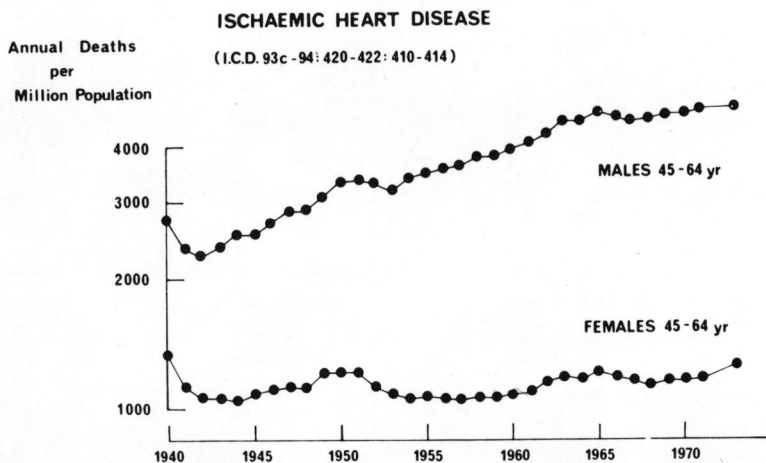

FIGURE 19. Deaths from hypertensive, cerebrovascular, and ischaemic disease, 1940–1973. (Data from Registrar-General's Statistical Reviews of England and Wales.) The International Classification of Diseases (ICD) was revised at the times indicated; the ICD categories used are shown. Death rates, adjusted for change in the age and sex distribution in the population, have been extracted for males and females of age 45–64 yrs.

adaptive change, and the possibility of beneficial adaptive changes being of general importance for the strategy of treatment of other chronic degenerative disease.

ACKNOWLEDGMENTS

We were indebted for help during the Jubilee Exhibition by Dr. J.M. Armstrong, Dr. B.N.C. Prichard and Dr. D.G. Davey, and to the Wellcome Foundation Limited for assistance with the display.